# A Salute to American Patriot Warriors

William L. Cathcart, Ph.D.

A Salute to American Patriot Warriors

By William L. Cathcart, Ph.D.

Please note the information contained within this book is for educational purposes only. All effort has been made to ensure that the content provided is accurate and helpful for our readers at publishing time. However, this is not an exhaustive treatment of the subjects. No liability is assumed for losses or damages due to the information provided.

ISBN: 978-1-7374922-2-1

Printed in the United States of America

# DEDICATION

This book is in honor of all the courageous men and women of America who, through generations, have chosen to serve, whether on Active-Duty, with the Reserve, or the National Guard, and if need be, remain fully dedicated to defending the sovereignty, security, and the freedom of our great nation.

CONTENTS

Acknowledgements

# ACKNOWLEDGMENTS

It is with grateful thanks to the warriors included within this volume, for taking their time to share with me memorable aspects of their respective stellar military service to America. All of these impressive military officers have either served, or lived, in the Savannah, GA area.

Sincerest thanks and appreciation go to my incredible wife, Dr. Julie Olsen (MBA & DPA), who despite the pressure of her own professional responsibilities, spent hours of her time and expertise making certain that the prepared elements for this book successfully reached publication. Quite simply, this project would not have been possible without her diligent encouragement and assistance, for which I will forever be thankful.

# A Salute to American Patriot Warriors

A Salute to American Patriot Warriors

# CHAPTER 1

## From Young Immigrant to West Point Graduate
## to Career Combat Engineer
## Colonel Jose L. Aguilar
## United States Army (Ret.)

Born in Mexico, young Jose Aguilar went to live with his aunt in Dickinson, Texas from age 9 on, and attended Catholic, then public schools, there. Upon graduation from high school, Aguilar had some important decisions to make. For starters: What's next? He had gone through the Air Force JROTC program at his high school, so perhaps continue on with that track to become an Air Force officer. But then a friend, one grade ahead, started talking about

West Point. These two possibilities now centered within his mind, as he began thinking more seriously about colleges. Aguilar was a long-time participant in martial arts, and some of his older friends and mentors from that tight-knit group began advising him on various higher education possibilities and, especially, as they were quick to remind, the costs of attending!

Contemplating all that he had absorbed on his own, and from conversations with friends, about which post-high school direction to take, Aguilar finally boiled it down to five key choices. His first & second would be attending a military academy, either Air Force or Army. His third option was to attend in state Texas A&M. Fourth, would be to attend, also in-state, the University of Houston. The fifth and last on his "what if" list was a non-university back-up possibility, something completely different: attending the state's police academy near Austin. He had a good friend in his martial arts group who was also an experienced police officer. So, he had gone to visit him, on the job, "to see what the possibilities were," remembered Aguilar.

Following that law enforcement exposure, his thoughts began turning back to an academic path. He realized, early on, that if he went to Texas A&M or Houston, he would have to work to handle the costs, whereas attending one of the academies, were he to be accepted, would be cost-free. Despite his high school years in Air Force JROTC, after considerable thought, he decided that attending West Point would definitely be his first choice. Now, all he had to do, was get in! A process, as he quickly discovered, that was considerably more involved than that of more regular university admission.

As he recalled: "It involved, first and huge, a Congressional nomination, along then with mental, physical, and dental evaluations, and the normal teacher (and/or employer) written recommendations, all were a part of the overall application process." His high school academics were strong (approximately 3.5/4.0), quite impressive enough, especially when remembering that he had to learn English when he moved here from Mexico! Meaning that he had started out in Texas schools one grade level behind, to help compensate for his language deficit. Additionally, with academics, he had progressed so well, that by high school, he was taking honors classes in the subjects that we today identify as STEM, "in order to get the best education possible."

Initiating the application requirements, he scheduled his mental and physical testing, which he successfully passed. Aguilar then outlined the three remaining requirements ("challenges") he faced, in order to successfully complete the full West Point application process. The first challenge, of course, was obtaining that required Congressional nomination, a task initially hampered by not knowing anyone, at that time, in federal political office. Later, while talking with friends, he learned that then-Texas Congressman Jack Brooks was due to speak soon at Galveston's County Park (Walter Hall Park). Aguilar decided that his goal would be to attend that speech and, when finished, approach the Congressman to request an Academy nomination. All went as planned. Congressman Brooks was very accommodating, "linking me up with one of his aides, a really nice young lady, who then walked me through the whole process, so that my paperwork would be complete," said Aguilar. That impactful encounter started the West Point Express rolling.

His second major personal challenge had to do with dental. Unbeknownst to him at the time, he discovered that he had some serious issues, all resulting from a chipped tooth, which had likely occurred many years before, during a playground fight. That chipped tooth had never been an issue, nor had it been closely looked at prior, resulting, now, in an infection found to be running through the tooth, down into the root canal area, and on down into the chin bone! This would result in the need for dental surgery, costly stuff, and a tough prospect for this potential West Point selectee, whose financial resources at that point were, needless to say, quite limited (understatement!). Where to get ideas and direction? It would be his long-time martial arts teacher who provided him with the needed 'pull'.

In fact, his karate instructor, whom he'd been with since age 13, had a couple of adult students who were dentists (what are the chances?). So, his instructor arranged for Aguilar to see one of them for his initial examination. That karate-student dentist took a look and said: "I can take care of just about everything, but the surgical part needs to be done by a specialist. I've got a friend and I'll ask him to see you." And, with that, the incredibly fortunate karate connections just kept going! Thanks to this understanding karate-connected dentist, the only thing he charged fellow-student Aguilar, both for his initial exam and procedural work, was the time spent by his dental assistant! All of this dentist's professional skill and effort to correct what he could of Aguilar's teeth was done at no charge. And his generous assistance for this impressive young man didn't stop there. He then referred Aguilar to a close friend and colleague, a surgical dentist, who took care of all that extensive deep infection issue, and, while not doing so completely free, as Aguilar recalls, his charge was very minimal, due no doubt as a favor to his dental colleague. And, also, quite probably because he knew that this

3

bright young man, a legal immigrant, was working very hard to gain admission to West Point, in order to become a stand-out officer in the United States Army.

So, that meant one challenge now completed (dental), one pending (Congressional appointment), and the last, a very big one indeed: Citizenship. Although here in America from an early age, at that time, he had neither obtained, nor apparently needed formal U.S. citizenship. But now, there's no way, of course, that he'd be entering West Point without it.

So, it was off to the nearest INS office (Immigration & Naturalization Service, back then). The folks there were, again, quite helpful, outlining what he needed to do for his citizenship test, and to that end, gave him the necessary study guide. Although he was not quite yet 18-years-of-age, those good folks decided to move ahead and allow him to take the required U.S. citizenship test, but they would not be releasing his score until his actual 18th birthday. He took the test. And on his actual birth date, he returned to INS, got his test score, and learned that he had passed. Next step: the scheduling of his official swearing in. Big problem. "Turned out, the schedule for my swearing-in was a one-year wait!" recalled Aguilar. Upon hearing that, as thoughts (and disappointment) raced through his mind, the realization hit him that he would have to do a year at a community college first, since there was now no way that he could complete his citizenship in time for that year's West Point enrollment.

And then, once again, fate intervened on his behalf. Seems there was a West Point recruiter in town (Colonel Jackson, as he remembers) working to convince top-prospect local high school seniors to consider the Academy. The West Point admissions office must have asked him to follow-up with applicant Aguilar, since virtually all his requirements but two had been met, but they'd heard nothing further from him. So, as Aguilar recalls, the Colonel called him on a Thursday. He brought the recruiter up to date on all that he had completed, but that in order to fully complete his citizenship requirement, the actual swearing-in, he was now facing a one-year wait! Shortly thereafter, they concluded their call.

Later that same day, the Colonel called back. "I have no idea what he did," said Aguilar, "but that Saturday, just two days later, I found myself standing on a huge football field in Houston for my swearing-in ceremony!" Thank you, Colonel Jackson!

As soon as that was all official, he sent the citizenship certification to West Point. Then he contacted Texas Congressman Brooks' office and sent the citizenship notice to him, as well. It wasn't but a week or two later that Aguilar received word back from the Congressman that he was to be one of his nominated candidates! A short time later, Aguilar then finally received word from West Point that he had been approved, "if he chose to accept the appointment," which of course he did, with zero hesitation! "So, at that point, I'm, now in!" recalled a greatly relieved "Cadet Candidate" Aguilar. It must be noted, here, that he had worked toward this acceptance goal all during his high school junior and senior years. This whole process wasn't days or weeks. It was a comprehensive, non-stop, many months-long effort.

And that's an important key to emphasize. Recalling that this whole, lengthy, winding application road began with an initial encounter with that Texas Congressman following his speech in a nearby community park. That whole process clearly demonstrated personal initiative on Aguilar's part, followed by the continued persistence needed to make it through all those goal-directed hoops. Did he have important help, and right-time/right-place good fortune along the way? Yes, indeed he did. But much of that would not have happened, had he not continually demonstrated that he was both worthy of, and quite serious about, West Point admission. With his solid academic achievement and his demonstrated personal character in tow, he was determined to get there, blessed with, and driven by, the initiative, discipline, dedication, persistence, and the follow-through needed to make his academic (and career) goal happen. Those are fundamentally important qualities that can determine initial & future career opportunities and achievement, whether in the military or in the civilian work force.

Moving forward, what then, looking back, turned out to be his most vivid memories of those mentally and physically demanding West Point years (1987-1991)? Following his arrival on campus, due to the rigorous First Year requirements, restrictions and expectations, he remembers "kids were dropping out left and right." Recalling Aguilar's own demanding path to West Point acceptance, there was no question that he was there to stay. Said he: "Unless they carry me out of here on a stretcher, I'm not leaving this place! I had made a commitment, yes, to the nation, but also to my family. I was not going to go back to my home and tell my dad that I had quit." By comparison, he did recall that more than a few of his beginning classmates had apparently relied on their parents to assist with, or handle outright, the complex application process. "Turns out, those students had no skin in the game, so that when it all became difficult and demanding, and, no question, it was (!), they just decided to leave."

Other key memories: "There is no other place, that I've ever been a part of, that gives you everything you need to excel, both physically, mentally, and spiritually, everything. There are great people you're surrounded by. You're given time for worship (very important to him). The library there contains everything you'd ever need academically. The instructors are great. And as instructors at West Point, through the ages have said, most of the history they teach there was made by people they taught! Also, you have the best physical training facilities, and you're afforded the opportunity to use them for your own betterment. There is just no better place for all-around self-improvement that I know of than West Point. And, you don't have to worry about funding. Everything's given to you," said Aguilar.

However, he did say, speaking to what some people may see as a negative; "There is NO free time for anything else. Your time is filled. You are improving yourself, whether you want to or not! Oh, and talk about meritocracy. Back in high school, you were constantly being judged by the clothes you wore, the things you did, your family's reputation, etc. But at the Academy, you quickly learn that everyone's on the same page. Same background. You are judged on what you do, or don't do, there. Honestly, to me, that was a pretty darn good environment. They continually challenged you," remembered Aguilar.

As for his academic preparation back at Dickinson High School, he felt that he was well prepared in the math and science areas, where he had taken honors classes and had excelled. But the Academy stressed well-rounded achievement in every aspect of Army officer preparation. So, when it came to subjects like history, English, and other social science/humanities courses at West Point, Aguilar felt that "to say I was in the bottom half, was a compliment to me! I had to work very hard in those subjects."

It was at about that time when computers were being introduced, both there and throughout American government, businesses, and society in general. He recalled that "ours was the first class to be computer-based on day one. And here's what really helped me with that. Back in grade school, a friend convinced me to take typing. Besides trying to find the power button on that thing (computer), if I hadn't taken typing, that would've been really tough, since, from then on, everything was required to be submitted on the computer, as they were then making that complete transition. And all of that was difficult for me," said Aguilar. For 1987 computer perspective: Microsoft released 'Word-3' and introduced 'Microsoft Works' that year. Also: The Mac SE was introduced and priced at $2,900 (in 1987 dollars!). And that same year, IBM released its first laptop, featuring dual 3.5" floppy-disk storage. It weighed 13 pounds (good for added strength conditioning!).

When asked whether he considered himself to have been well prepared, overall, when he showed up at West Point, his response was an emphatic: "NO! It was very difficult. Not only looking at the academics, but the rigor of the entire program. Academics was only one part of it. The rest of it was memorizing things; having to work hard from very early morning to very late at night, and still continuing to study. Your every moment had a purpose. And that purpose is to better develop yourself. Even when you take a little bit of down-time to maintain your sanity, it has a purpose, and that purpose, no surprise, is being able to maintain your sanity, so that you can continue to do the things you need to do," recalled Aguilar.

At the Academy, every cadet must be involved in some form of athletics, either a school team or intramural. Aguilar's sport was martial arts. He entered West Point as a Black Belt in karate, and so he joined that team (and became its co-captain during his senior year). He and his fellow karate cadets competed against Ivy League schools, the Air Force Academy, and others. "When I spoke about maintaining sanity," said Aguilar, "karate was one of the things I excelled at. Like other things in life, I had dedicated the necessary time, energy, and effort, over many years, to be good at it. And it was something that I really, really enjoyed. So, I think that activity (karate), coupled with going to Mass every Sunday, were the things right from the beginning, before developing relationships with classmates, that really gave me the foundation and the motivation to continue, as demanding as it was, at West Point."

As he prepared for graduation and commissioning (1991), along with final exams, Aguilar needed to submit his branch requests. His list of preferences came down to three: Aviation, Infantry, and Engineers, in that order. After pondering (just a fancy, almost-second-lieutenant, word for thinking about), he decided against aviation because, after four tough years at the Academy, yet another year of schooling just didn't appeal to him. Rather, he really just wanted to start contributing right away. So, then: (1) Infantry and (2) Engineers, became his adjusted preferences. That is until one night, very close to the submission cut-off, while studying for an engineering exam, he began to think about all of the coursework time he'd spent in that major. Then and there, he concluded that it would be foolish not to pursue the Engineer branch as his first choice. He rationalized his decision by recalling that Army engineers spend a lot of time in the field, like the infantry, so engineering might well be the best of both worlds. "So, at the last minute, I withdrew Infantry, and selected Engineers as my first preference…and I got it!"

After branch, came selection of location. To which post or unit did he hope to be assigned? "Frankly, there wasn't much left on the board when I got a chance to pick," recalled Aguilar. "So, I ended up picking Fort Sill, Oklahoma. Partly because I didn't want to go to Korea, which did not appeal to me. And there were a couple of other places I wasn't excited about. But Fort Sill was close to Texas, so on weekends, I could go visit family. And quite honestly, thinking back to those days and decisions, my goal was to do my five-year commitment, followed by three-years of ready reserve, and then I'm out. At the time (1991), that was my thought," he remembered. But how times would change for this Second Lieutenant, and all for the good.

Before heading to Fort Sill (which ended up being Fort Carson instead, because his assigned engineer unit was then scheduled to move to the latter), Aguilar was, instead, off to the Engineer Officer Basic Course at Fort Leonard Wood (MO). Of interest to Aguilar there, far more than the instruction, was meeting fellow engineering student, 2LT Suzanne Calahong, who would become his wife. Even more interesting and serendipitous, they had both graduated from West Point, in the same Class Year, but had never actually met while at the Academy! But first things first. Following the Leonard Wood course, Aguilar was off to Ranger School at Fort Benning (GA), in January 1992, just six months after West Point graduation.

Army Ranger Training: A key memory he had from the rigors of that Special Operations instruction sequence was that it teaches you about your limits. "And then after you understand what you think your limits are, it teaches you that you can actually go further," recalled Aguilar. But through it all, the hard, long, cold, sleep & food deprived, physical and mental challenges, he concluded: "It also teaches you something else that I will always use. And that is, when you are part of a team, whether you're the leader or led, you need to give it your all to accomplish the mission. And that's because, pretty soon, people figure out whether you just shine when you're put in a leadership position (termed 'Spotlight Ranger'), or whether you're a team player all the way through." This lesson has direct military career applications and relevance regardless of branch or specialty. "Within the Army, you go from a leadership position to a staff position, and you need to perform well in all of those assignments. Not just coast, because you're in a staff position, and wait until you're a leader to really work hard. Because, normally, if you're working hard in a staff position, that's making a leadership position easier for someone else. So, that's a huge lesson that I learned," he said.

There are several training phases for future Rangers (three now, but four when Aguilar was going through). The first, at Fort Benning, focused on squad level basics and missions. Phase two was mountain and focused on platoon level operations. Three was desert. And four was the "swamp" phase in Florida. Fresh back from leadership in Gulf War I, General Barry McCaffrey was his class graduation speaker, marking the conclusion of Aguilar's Ranger experience, which finished up back at Fort Benning. Sixty-seven days of hard, humbling training later, Aguilar proudly wore his new Ranger Tab, as he headed now, finally, to Fort Carson (CO).

There he joined the 299th Engineer Battalion as a Sapper Platoon Leader. Or almost. His unit had not quite yet arrived from Fort Sill, so he and two others worked to make preparations for that anticipated relocation. The delay also gave him time to become thoroughly familiar with Carson's facilities and its ranges. "I knew down-range Fort Carson probably better than those young soldiers who had been there for three or four years, because I spent so much time in the field, which really helped me out when my platoon and battalion arrived. I knew that place like the back of my hand. I was the expert, instead of showing up as the fresh lieutenant who doesn't know anything!" recalled Aguilar. He served there as a Platoon Leader (2 different platoons), Assistant Operations Officer, as Battalion Maintenance Officer, and as Civil Engineer for the 52nd Engineer Battalion, now at the rank of Captain. He and his now-wife, Suzanne, also an Army Captain, ended up spending about four years in service at Fort Carson (1992-96).

Next came the Engineer Officer Advanced Course at Fort Leonard Wood (MO), where Aguilar had previously spent a month (1993) completing the Sapper Leader Course. The Sapper Course experience, he remembers: "was just as rigorous, but not as long as Ranger School, and there were more specific mental elements (demolition calculations, identification of mines, etc.)." As a part of their training, he and his fellow students also learned how to kill, clean, and cook a chicken and rabbit, if needed for protein, out in the field. Handy to know, no doubt, although perhaps surprisingly, he has yet to demonstrate that skill on the back patio for his family at home. They no doubt thank you for your restraint, sir.

*CPT Jose Aguilar (Center Standing),*
*Commander, Bravo Company,*
*54th Engineer Battalion (1998-1999),*
*Bamberg.*

From the Engineer Advanced Course, he and his family were off to his first overseas assignment: Bamberg, Germany (1997-2000), where he began as a staff plans officer, followed up by a brief time as the acting operations officer (a Major's position), before assuming command of Bravo Company of the 54th Engineer Battalion. While there, also serving, his wife, Captain Suzanne Aguilar, made the decision to leave service (1999) to become a full-time Mom for their growing family.

Three years later and back in the States, Aguilar's next assignment (2000-2002) was as an Observer-Controller at the Army's National Training Center, at Fort Irwin, California. There he served as a Maintenance Trainer, then as the Engineer for the Airborne "Tarantula" Team, and finally, as a Company Trainer. The NTC provides the most realistic combat training through force-on-force (lasers) and live fire operations at the Brigade level. He and his fellow Observer-Controllers focused on teaching, coaching, and mentoring units while honing their own skills as practitioners of the profession of arms. During his time at NTC, Aguilar pinned on the rank of Major, and in 2002, he was selected to spend a year (2002-2003) studying at the Army's Command and General Staff College, located at Fort Leavenworth, Kansas. "That school teaches you about brigade-level to Army-level operations. So, from a tactics perspective, now you're honing your skills to be able to operate as a staff officer, or a leader, within the broader and bigger organizations in the Army," said Aguilar.

Prior to CGSC course completion, he had interviewed for several next assignment locations, with his first choice being Fort Riley (KS), a comparatively convenient in-state family move from Fort Leavenworth. And he got his first choice! He and his family would be at Fort Riley from 2003 until about 2008. There, he served as the Group Maintenance Officer for the 937th Engineer Group. Then as the Group S-3, before moving to the 1st Engineer Battalion, also as the S-3, where he would go on to become its Executive Officer. What followed during the Fort Riley stationing, would be two deployments to Iraq, one with the 937th Engineer Group (2003-2004), and one with the 1st Engineer Battalion (2006-2007).

He arrived for his first Iraq assignment in June 2003, "after the initial ground fight was complete. We stayed there to help train and stand up a new Iraqi Army. That was our mission," remembered Aguilar. Some heavy fighting did, however, continue, with

welcome invaluable combat help and force-strength supplied by the Third Infantry Division from Fort Stewart, Georgia.

His second Iraq deployment (2006-2007), with the 1st Engineer Battalion, was a focused combat route clearance mission in Multi-National Division North, and that was a 12-month tour, which turned into 15-months, prompted by the "Surge."

As 1st Engineer Battalion Executive Officer, for Major Aguilar, the "Surge" deployment was definitely the tougher of the two, resulting in the loss of two of his soldiers. "Two is too many to lose," he said, thinking back, "but with all of the (enemy) contact we had on nearly a daily basis, it was a miracle (that our deaths were limited to two), clearly assisted by the heavy training we did, and the great work that our junior leaders did there to protect their troops." Aguilar remembers going out with his troops once every couple of weeks to do "leadership circulation" on the ground, while engaging in those patrols with them. "But really, the guys who saw some hairy stuff were my platoon-level leaders and below, because those guys were out there every single day, sometimes a couple of times a day, or if not, then on very long patrols, where they were under frequent enemy contact, and having to do whatever it took to keep those routes open for everyone else," remembered Aguilar. And those routes simply had to be kept open, not only for the military, but in order to protect the surrounding local populace, as well, because "IED's are indiscriminate," he recalled, from direct observation.

*MAJ Jose Aguilar (left) with
MG Eaton,
in the middle of the desert,
training the new Iraqi Army.*

And some of his personal thoughts, reflecting back on those two Sergeants who had been lost. "There's a certain amount of guilt within the leadership when you lose someone. The very first question you ask yourself, after you conduct your 'battle drill,' making sure the wounded are evacuated, and there's no danger to others, is what did I not do right, that caused us to lose someone? I think that's a natural thing we all go through. Pretty soon, you have to remind yourself to figure out how to grieve very quickly, and then move on, because there are six-hundred other soldiers that you're responsible for, many of whom are right now actually out on patrol," said Aguilar.

Fortunately, he never had to medically "evac" one of his own soldiers, but he did have to do so with a couple of local Iraqi civilians who got hurt near-by. The very next morning, just after losing one of their NCO's, Aguilar went out on mounted patrol. Traveling with his team, they spotted a vehicle, sitting off to the side of the road, that just looked suspicious. It was similar to one that had exploded the previous day. The patrol "executed a battle drill," where they began setting-up security on the far side of this four-lane divided highway. At that point, without any sound or warning, that parked red sedan just blew up. "It basically disappeared; the vehicle was no longer there," remembered Aguilar. They immediately radioed up to their lead vehicles to be certain all personnel were OK. Other than one flat tire, all was well with his troops. Enemy insurgents appeared to be nowhere around, having apparently detonated the explosion remotely. But although our soldiers were fine, a couple of approaching locals soon were not.

*MAJ Jose Aguilar (right),*
*near the Iranian Border, at an Iraqi Training Base.*

The insurgents had positioned some mines where they thought Aguilar's troops might be positioned in relation to that bomb-laden red sedan. Before long, a civilian tractor-trailer approached, with the driver instinctively moving to the side of the road to get out of the way of the vehicle explosion he'd just witnessed. In so doing, the truck hit a mine which blew off the entire right front axle. Riding in the cab of that truck, with the older male driver, was his nephew. Both were hurt from the mine blast, but not critically. Nonetheless, the patrol had to have both individuals evacuated for medical treatment. It all could have been worse for Aguilar's men, but good fortune was riding with them on that patrol.

And about the enemy insurgents his soldiers were then facing in that sector of Iraq. Those enemy forces "were a combination of folks from the previous regime, along with some disillusioned others, with no way to make a living, so others would pay them to do those kinds of things (destructive acts). Not making excuses for that second group, because they, too, hurt us. And then there was also a combination of Iranian influences coming in there and assisting those disillusioned. That's because Iran had everything to gain by getting involved and trying to prevent the new government (from succeeding) and, thus, preventing us from achieving peace in the area. It was all just a combination of folks. In some cases, it may have even just been kids who didn't have anything better to do, so they were making a couple of dollars or making a name for themselves," remembered Aguilar. Make a name or make a buck!

Looking back at his second Iraq deployment, Aguilar was asked what he most remembered about that experience. Said he: "I was just so impressed with our junior leaders. The reason is the discipline and the restraint they showed which was phenomenal, because there were some really bad people out there trying to hurt us." As but one example of both discipline and restraint, Aguilar spoke of the times when enemy individuals would be captured, and then our junior leaders had to travel to Baghdad to testify at their trials. He recalled that too many times, for a variety of reasons, those insurgents, known by our soldiers to be guilty of those crimes, were simply let go. What that meant was that not only was a patrol(s) back in the field going out short those two soldiers, but also that those same junior leaders had been sent away on a wasted trip. Our NCOs were trained to treat their prisoners with respect, "as much as you can say that in warfare. Respect for their capabilities and respect for them as human beings," said Aguilar. Laudable, especially, when you remember that the insurgent's only goal was to kill or seriously wound American troops.

The second thing he remembered most about that time, centered more on his own job as an XO and the functions of the Battalion, "enabling those that are on the ground to have everything they need in order to be, and execute, 'over-matched' against the enemy." In that regard, he recalled anxiously awaiting their shipment of Mine-Resistant Vehicles (rolled-steel, V-shaped deflecting bottoms) to arrive. And that was because "every day when our soldiers were out in those far safer vehicles, was a day I worried less. So, we were tracking their (vehicles) arrival and trying to get our units outfitted with the best optics, the best protection, the best jammers, with the best everything, to make sure that they had the very best chance out there facing danger," said Aguilar. "From the battalion all the way up to the national level, and the American people, the efforts we went through to ensure that our soldiers had the upper hand in combat was phenomenal. From the investments we made in those superior protective vehicles, to the medical attention that was given all the way back to hospitals in the States, and how folks survived who would not have even ten years earlier. In my view, that backing, and effort, was all just terrific."

In January of 2008, Major Aguilar & family headed to Portland, (OR), as the Deputy District Commander and Deputy District Engineer for the Portland District, United States Army Corps of Engineers. When he was still in Iraq, and beginning to think about his next assignment, he remembered that his Branch Officer (assignments) had called him, saying: "I've got a couple of options for you.  You can either come back home, high-five your wife, and go back to Iraq, or you can go to Korea, or you can go be on an observer-controller team that trains and travels all the time (based out of Fort Leavenworth)."

Still in Iraq and on his second and long (15 month) deployment, Aguilar remembers his response was as follows: "Listen, I probably don't need to tell my wife what options you just gave me, so let me see what I can find and I'll get back with you!" Given those less-than-attractive options, both for him, and especially for his family, this became the first time in his career that he realized he could actually have at least some say in where he wanted to go next.  His wife, Suzanne, simply asked him to be with the family for at least a year.  The branch assigners are always up to their elbows in work, trying to fill open positions, or as Aguilar put it, "putting butts in seats."  Further, he said: "I respect what they do and would never want that job, but I had been away too long."  He felt he could actually try to help the process along by hopefully finding a compatible open slot by himself.  And, in this case, it worked!  He did some serious looking around, primarily by contacting a number of Army friends for any job leads they might have.  Then, at some point, Suzanne suggested she'd never been to the Pacific Northwest, so perhaps he could find something in that region?  For that possibility, Aguilar then turned his search to the Corp of Engineers for any potential openings they might have in their Northwest region.  As (well deserved) good fortune smiled down upon him once again, he located a soon-to-be-open CoE Deputy Commander and Deputy District Engineer position. He interviewed for it, got it, and the Aguilars were then happily on their way to, not Iraq, not South Korea, but Portland, Oregon!

Now, before moving into a discussion of his military-civilian Corps of Engineers experience, and after his having successfully completed two Iraq engineer deployments, from his unique experiential perspective, Major Aguilar was asked to go back several years in time and focus specifically on, and fully explain, the role of combat engineers within the total Army engineer compliment.

Combat engineers, he said, provide mobility, counter-mobility, and survivability support for the warfighters. Aguilar described the specifics of those combat requirements as follows: "For mobility, if there's an obstacle, you end up bypassing or breaching it to allow the forces to continue forward. If there's a trail that needs to be cut in order to allow forces to move ahead to the desired destination, they would make that trail with the appropriate equipment. With counter-mobility, if you're on the defense, then you create obstacles to prevent the enemy from just attacking and reaching you faster, with nothing in their way. Those imposed obstacles, then, are to allow the maneuver forces to have more time to engage the enemy and destroy them or prevent them from going to an area where you may have weaknesses. Lastly, survivability would be things like digging a fighting position for our tanks, so that they would only come up into partial view when they engage the enemy. Or, also, to protect our soldiers, by digging positions for them," he said.

Along with the combat engineer, there are also Army construction engineers, whose functions include building infrastructure, roads, and airfields, among other essential projects to assist our fighting forces. In some cases, secondarily to the needs of the fight, they may also work to fulfill some construction project needs that would help the local populace, as well. On this subject, it's interesting to note that while Captains Aguilar were both stationed at Fort Carson, Jose Aguilar was a combat engineer, while Suzanne Aguilar served on the construction engineer side. Both specialties are critical to the success of our fighting force missions. Besides combat and construction designations, additional Army engineer functions include diving, firefighting, map making, and terrain analysis (pre-patrol advice to soldiers on specific ground formations and elevations they will encounter).

But now, out there, in the fight itself, what is the riskiest thing that combat engineers might have to face? "In my view," said Aguilar, "that would be the task of breaching. Doctrinally, it's a combined arms effort. You're up there with armor and infantry forces, lead elements, and you're having to move forward, under fire, to find and bypass or breach whatever the obstacle is (identify via prior reconnaissance, or worst case, just bumping into it!), in order to allow our force to then penetrate through it, and assault to destroy the enemy. What that means is, you're already in an area that is dangerous, like, a mine field, for instance, plus you're out there in front, and quite probably being fired upon. Now, if you have some pretty good maneuver guys, they're protecting you, and destroying them or at least making the enemy troops keep their heads down. But, regardless, indirect fire is, likely, still coming in on you. So, from my experience, breaching is the most dangerous thing our combat engineers have to do," he said.

Once the obstacle is cleared, the engineers then create the initial penetration and mark a lane to the far side, then secure it, making way for the assault force, which then "punches" through, while our support force continues to keep the enemy occupied, as the assaulters close with and destroy the enemy, clearing the far side of the objective. Then everyone on the remaining team moves through, consolidating and reorganizing on the mission objective. The next concern: Preparing while waiting for the anticipated counterattack, as the enemy attempts to take back that position. At which point the engineers may now be in the role of counter-mobility.

"However, if you notice what happened in Iraq and Afghanistan," said Aguilar, "engineer units were, for the most part, functioning on their own to execute missions, out front, where, in that instance, they searched for IED's and cleared them. The reason we were able to do that was because the enemy was not one that could overwhelm and overpower an engineer platoon. Our engineer platoons could hold their own, with the firepower we had, utilizing the combined arms team. Very often, the first element of that combined team focus was aviation, used to bring fires down on the enemy from above. And then, another maneuver force, like an infantry or armor company, that is, if the enemy's force was big enough, would be called in to make contact with, and destroy, the enemy, while the engineers then completed their mission. That's why, in both Iraq and Afghanistan, when necessary, you saw Army engineers functioning on their own because, number one, the enemy wasn't big enough to provide a threat to annihilate an engineer platoon, and number two, there wasn't enough coalition combat power to go around."

Aguilar explained further: "Enemy forces in Iraq and Afghanistan could not afford to be that large because, if they left a big enough 'footprint,' we could more easily find them, fix them, and destroy them, either by indirect fire, Air Force fires, or with one of our ground units. What they chose to do, instead, was to function more in small teams, or even as individuals, and so most of the time, we didn't really need someone protecting our engineers, since they usually had enough combat power to find and destroy the enemy on their own. We always wanted 'over-match.' We never want to put any of our military members out there in a position where, unexpectedly, it becomes our own guys who end up over-matched! Now, if there are warnings and indicators of some large element that could threaten one of ours, then we employ additional ground elements or fires. And, yes, that did happen! In which case we needed to employ the air or ground fires that could effectively attrite the enemy threat, prior to them making contact with our forces, and then our forces can either handle the situation, or at least hold that position until reinforcements arrive," he said.

How did combat engineers actually locate mines and IED's? Replied Aguilar: "First, it's important to understand that our concern was not with the entire country of Iraq! We focused on things like lines of communication, and roads, within our specific area of responsibility. The key to informed route (road) clearance is 'change and detection.' We were using mounted patrols, traversing that terrain all the time, so that our soldiers (aided by optics) were able to observe things like a curve in the road that now doesn't look like it looked yesterday. Or the people in the little villages are not acting the way they were before. Or that car was not there on our last patrol; what is it doing there? Detecting change. That's the responsibility of those there on the ground to figure out. For instance, using that first example: Is that a new, but fake, curve in the road built to contain an IED? Does the dirt look fresh, disturbed, or discolored? It was our lead guys who were spotting those kinds of things, different objects, different surroundings, any changes. Their observations were then confirmed by using the very latest equipment, such as optical aids, robots, and the other technology that makes our Army of today so lethal," said Aguilar.

Regarding the critical need for mine and IED detection, Aguilar continued: "The second detection factor (after, observed 'change') is that back at the Command Post, there's information that's being gathered from all kinds of sources, that tells you what kinds of tactics, techniques, and procedures certain enemy elements are using, so that, you can try to predict (predictive analysis/historic trends) where hazards may exist or enemy lethal tactics may be migrating across the battlefield."

To give you an idea of the success, it was just a few months after we were in theatre that our intelligence 'shop' in the battalion was able to gather information from all echelons, and then tell our patrol before departing, for example, that sometime between the hours of x and y, you will have an individual placing an IED here, and it'll be this kind of IED (!). So, now you're in the enemy's cycle, and what you're then able to do is drop off a team that's going to observe that intersection, or that area, and when the enemy individual(s) arrive, we report back, and since we already have a team in place, we can take appropriate action. That collective intelligence allowed us to literally get in front of the enemy's decision cycle, and now we're on the offense, rather than having to play defense. There's an Army saying:  Intelligence drives operations; operations drives logistics and it's a cycle, not linear.  And, yes, offense is better than defense!" said Aguilar.

Summarizing his detailed description of combat engineers and their functions: "Even when you're out going into the middle of nowhere, you would have combat engineers along in case you find an obstacle or explosive, so they could go ahead and clear it, in order to allow the force to continue to their objective."

Back now to Major Aguilar's new Deputy District Engineer experience in Portland, Oregon. He was asked why pursue and gain the position in Portland, other than the fact that it was, at that time, preferable for both he and his family, rather than, say, going to South Korea or spending yet more time in Iraq?  "Because I knew that, as an engineer, I needed to understand how things were done at the national level in order to progress in my career.  During my first time in Iraq, I was impressed by a Colonel who joined our group, and who had previously been with the Corps of Engineers.  When we had to develop the program for the new Iraqi Army, and then explain the funding needed for facilities, equipment, etc., he (Colonel) was able to put the project request into 'Congressional language;' its scope, justification, and estimate of total costs, for eventual submission to the Administration and then, Congress, in order to obtain that needed funding. You need both the authority, and then the monies, to be able to do the assigned mission (in this case, the enormous task of standing up a new Iraqi Army). I was impressed with his ability to communicate so effectively at the national (civilian) level," recalled Aguilar.

What, then, was his primary career value take-away, from his time there with the Corps in Portland? "Frankly, the value to me was, I think, much larger than the value I provided to the organization. Because when I left, I'd learned so much, and I felt like I really hadn't contributed a lot. They spent a great deal of time getting me up to speed to understand how the Corps of Engineers operates. Also getting me to thoroughly understand, perhaps for the first time, that 'everything costs!' For instance, if we had a meeting with, say ten people, at a time & a salary rate of so-much per person attending, I now had to think, was that meeting really worth the collective cost? And, if not, why? And what could we do to change that? Because the Corps of Engineers runs itself like a business. Everything costs. So, whatever actions you take, will have an impact on expenses," said Aguilar. By the way, it was really refreshing to hear that the CoE does its very best to operate like the private sector, where entities must live by the bottom line, and where there is no Treasury printing press for back-up and bail-out! Government (federal/state/local) organizations don't always seem to operate within that reality. Understatement, perhaps?

*LTC Jose Aguilar, wife Suzanne, and family at his promotion ceremony in Portland, Oregon with the Corps of Engineers, 2009.*

And then, after 'everything costs," what was the second thing of value that Aguilar learned during his time in Portland? "I learned how to work with the civilian work force," he said. It's not that they're any less motivated than the military, they're just motivated in a different way. They're just as dedicated, they're just as committed to the mission. But with the military, if required in a crisis, you can work your troops 24/7. Not so, of course, with civilians. You need to give them very specific guidance (vs. 'orders'!), and you need to be sure you pay them for any overtime, along with assorted other civilian workforce policies. For instance, with the military, you can give soldiers a paid day off. With civilians, you can't just give them a day off with pay! That's not the way the system works. So, it took me a little bit to figure out that the rules were quite different!"

"Finally," said Aguilar, "it (time in Portland) gave me a view into the broader perspective of how things are done in our government(s). As a part of my responsibilities, I was helping my commander brief Congressional delegations, as well as meeting with state and local (city/county) governments (about things needing to be done in their respective areas). We were meeting with Native American tribal members (Pacific Northwest tribes), who had treaties with the U.S., and we were trying to work through opportunities and challenges within those commitments. We met with environmental groups. Then there was evaluating proposals submitted by contractors to determine the best value for the government. All of this, again, allowed me to see the broader federal/state/local perspective. And then, internal to the district, it showed me that you also have to be continually concerned about public affairs. To be certain that everything is well coordinated among the District's lawyers, engineers, etc., and so that we can be publicly, as well as administratively, successful with the missions assigned to us by the Congress and the Administration," said Aguilar, concluding his key recollections from his first, of ultimately two, assignments with the Portland District.

Then, in 2009, now-Lieutenant-Colonel Aguilar moved into Battalion-level command, as he began his two-year tenure (2009-2011) as the Garrison Commander at Hunter Army Airfield in Savannah, Georgia. "The really cool thing about that assignment,"

LTC Aguilar recalled," was that I had just come from an assignment where I understood the broader perspective of working with civilians, along with making some decisions that have political implications, so that previous job (Corps of Engineers) really helped me to step into the Garrison Command, now more fully understanding that civilian employees make huge contributions to America's Army," said Aguilar.

Moving, now, into his thoughts about his Hunter (HAAF) Command experience: "The most rewarding thing about that assignment was the people. And the toughest thing about it was…the people!" (a knowing chuckle from the LTC!). Rewarding: "There were so many phenomenal folks, inside and outside of the gate, that helped me get the mission done there as the commander. Great, great people. As but one example, and an important one, Glenda Johnson (veteran garrison command executive administrative assistant) "who took me under her wing, being probably her 20th Garrison Commander, and she didn't use that as a power base to get something, but used it as an educational piece for me, to make sure that I started off running and making contributions, helping our community, both inside and outside of the gate. So, people were the favorite and the best thing about that assignment," remembered Aguilar.

But then, there were some challenges during his Hunter tenure. As he had stated above, some people were not, as it turned out, a good and proper fit. "Within any organization, there are bound to be a couple of bad apples," he said. "And unfortunately, toward the end of my command, I had no choice but to relieve two senior leaders. Now, we all have faults, as do I. Their huge fault, that led directly to their downfall, was that they believed that rank has its privileges, but without responsibility, or any accountability, to others. Therefore, they committed some offenses and because they were in positions of power, felt they could, continually, get away with it. They had to have known it was wrong. But they just felt like they were above the 'law,' because they were senior. Both of those cases were very, very difficult for me to take the action(s) necessary to ensure that each one was no longer a part of the team," recalled Aguilar.

Were there any further thoughts that LTC Aguilar wanted to share about his Garrison Commander time at Hunter? "I just can't say enough about how supportive that community was," he said, thinking back. "I was amazed. As an example, there was one Christmas when all of our kids had one, maybe two, presents from the local community. They brought in trailer-loads of presents. Bikes, you name it! Then another example that comes to mind was when Chatham Steel Corporation in Savannah (CEO Bert Tennenbaum) came in and donated time and money to make the Halloween House (on-post) just so much fun for our kids. I could give you examples of that across-the board. Another one, WTOC-TV, the types of things that organization did to help with getting our message out when something was happening (on-post), so that we could make sure that our side, too, was being communicated. So, from the Chamber, from the Rotary Club, from other local entities, everybody was just incredibly supportive," said Aguilar.

As a personal side note, and not a given for all military stationing locations, the residents of Savannah and surrounding communities have a very long history of exceptional pro-military commitment, with strong support, specifically, for the installations and units we're blessed to have within our region. The community, and many of its organizations, appreciated the openness, and the sincere desire to become involved outside the gate, demonstrated continually by LTC Aguilar. Those efforts, that friendliness, were reciprocated in kind by Savannah and the surrounding area. Needless to say, he made, and left, a very positive impression for the Army, for Hunter, and or himself, during his time here with us.

So, then, as he prepared to finish out his two-year command tour at Hunter Army Airfield, in 2011, LTC Aguilar interviewed for the position, and was selected to become the next Chief of Staff to the Commandant of the United States Engineer School at Fort Leonard Wood, Missouri. Following his time in command, working with a large number of military civilian employees, and after his previous west coast experience with the largely civilian workforce at the Corps of Engineers, Aguilar saw the Engineer School assignment as the chance "to go and 're-green' myself to what Army engineers were doing, what equipment we're bringing in, how we're training soldiers, etc. As the Chief of Staff, the Basic (engineer) Course, the Advanced Course, the Sapper Leader Course, and the Museum, all those things fell under the Commandant's Office, which enabled me to play a role in influencing some of those activities, which was pretty neat, actually, since I'd gone through all of that training," he said. Although he would have preferred to remain in that position for the normal two years, LTC Aguilar was competitively selected to attend the Army War College, and so, of necessity, his staff tour at Fort Leonard Wood was cut short, but for the very best of reasons, the prestigious appointment to the Army's senior officer school!

From 2012-2013, the LTC (and family) was in residence at the War College in Carlisle, Pennsylvania. "There, you're directed to read, and that becomes your primary focus, and to further develop yourself, kind of like at West Point. They put everything in front of you to better yourself, to think at a higher level. You are assigned to read some of the military classics, to go back and look at some things that were done, look at some after-action reviews for certain events and missions, and then pull one's thoughts together and write! My 'capstone' thesis paper was about the positive and negative impacts of water on security, here and across the world. Some of those impacts we're actually seeing take place now. What that experience (War College) actually did for me was to allow me to have a little bit more time with my family. Also, importantly, it allowed me the time to think, allowed me to read, and that enabled me to then think more deeply about my profession, and how it relates to the nation and the world, hopefully a secure and prosperous one," said Aguilar.

His time, study, and reflection in Carlisle resulted in his earning a Masters' Degree of Strategic Studies. Former Fort Stewart/HAAF Commanding General, MG Tony Cucolo, was the Army War College Commandant during Aguilar's time there. Coincidentally, he was also the CG during LTC Aguilar's time in command at Hunter AAF Garrison! Now retired from the Army, General Cucolo was back then, and still is today, an outstanding, dedicated, smart, and incredibly articulate leader.

Upon completing his Army War College assignment, which included his promotion to full Colonel (which MG Cucolo officiated), Aguilar felt the need to stabilize the family, since his daughter, Allison, would be entering her senior year and a move at that point would have put her in four different schools in four years(!). So, he applied for, and received, a one-year teaching position right there in Carlisle at the

War College. Or so he thought! Seems the Army had another plan for its new Colonel. While his family would, in fact, be staying that next year in Carlisle, Colonel Aguilar would not. As the needs of the Army prevailed, he would be spending that next year (2013-2014) deployed to Afghanistan as the Chief of Basing, for then Lieutenant General Milley, Commander of the ISAF Joint-Command. At that time, we were reducing the number of bases there, as we began reducing, as well, American force strength. "My role, then, was to orchestrate the draw-down of bases, coordinating those decisions with the Afghan Government and with our allied nations, including, of course, the U.S."

*LTC Jose Aguilar, (center), with representatives of Defense and Finance Ministries of Afghanistan.*

*LTC Jose Aguilar talking with our NATO team in Afghanistan.*

At just about the time, he was preparing to leave Afghanistan, he was selected to return to the Corp of Engineers in Portland, this time as District Commander (2014-2017). "That job was phenomenal because of all the things we were doing to conduct the civil works mission throughout Oregon, SW Washington, and a sliver of Northern California. The primary missions were to run the dams

(Bonneville, etc.) on the Columbia River to enable the transportation of goods, and to produce hydropower for the Pacific Northwest. A collateral mission, then, was dredging those rivers and harbors to help ensure the flow of commerce. Then, on the Willamette River, there was also some hydro-power production, while providing for recreation, as well as the environmental restoration of fisheries, and those type of things. And as before, interacting with tribes, Congressional members, and with local and state governments. It was just a great, great job," recalled Aguilar.

*COL Jose Aguilar, (center), presented the Portland District Colors by his team.*

Regarding his on-going need to meet and negotiate with regional Native American tribal representatives, the key from the Corps' standpoint was maintaining 'responsible development.' "As we develop certain areas for industry or for commerce, our goal was to ensure that it was not having a negative impact on the environment. Development vs. environment. Protection of the environment. So, with the tribes, as well as other concerned groups, what we had to do was demonstrate that there would be no adverse impact, or if there might be potential for such, what we were doing to effectively mitigate it, to every extent possible" he said.

His next tour, then, would become the final active-duty assignment in Colonel Aguilar's almost three-decade U.S. Army career. From 2017-2020, he would serve as the Deputy Director of Logistics and Engineering, and the Command Engineer, Headquarters, North American Aerospace Defense Command (NORAD) and United States Northern Command (USNORTHCOM), at Peterson Air Force Base in Colorado Springs, Colorado.

*COL Jose Aguilar in discussions with Admiral Padilla of Mexico's SEMAR (Navy) in order to develop capabilities through Theater Security Cooperation.*

NORTHCOM is one of only six American geographic combatant commands globally, meaning that Aguilar had been selected for a position of broader responsibility, at a higher level of service, than he had ever before experienced. While NORAD is a bi-national command, which answers to Canada's Prime Minister and the President of the United States and is focused on the Defense of North America. For both of the commands, said Aguilar: "It's about defending our homelands. Defending the U.S. and Canada. And then, also, coordinating with Mexico and the Bahamas for the defense of our southern borders and the Caribbean. Understanding that those threats are out there, and that we need to continue to be vigilant, realizing along with that, the need to develop technologies and weapons systems to counter our enemies, so that we can protect our way of life."

Another aspect of his operational reality and responsibility at this level is "the complexity of how our nations make defense investments, and how we need to have open discussions with our partners to be certain that they are investing in the right areas, to ensure the security and prosperity of our respective nations," said Aguilar.

*NORAD's Cheyenne Mountain Complex.*

Then there are the actions taken through NORTHCOM for the U.S. Department of Defense support of civil authorities: "Things like response to hurricanes and wildfires, as well as any other kind of natural and man-made disasters, always in support of a civilian lead federal agency (for instance, FEMA). It's very sobering to see that we get involved with all of that (aside from the nation's defense), when the other agencies of our federal government don't have the necessary resources in some cases to address every situation that might arise. It's at that point that they will reach out to DoD to come in and assist," he said. Aguilar spoke also of the work that NORTHCOM does with our partner countries in the area of security operations, to be certain that all is in the best interest of our collective nations and needs.

And, finally, on a personal note, he said: "It's just pretty neat to see that, from a stationing perspective, we've now come full circle, having started our Army career here in Colorado Springs (Fort Carson), and now finishing here as well. It's also special because our two daughters were born here (their son was born during their deployment to Germany). Suzanne and I started here with no kids at home, and we end here with no kids at home!"

*COL Jose Aguilar and his wife Suzanne at the Portland District Birthday Ball.*

The Colonel was then asked, through all of these years, what the Army had meant to him. "The Army's been my life! You could make the argument that I really don't exist outside of the Army. Although with retirement coming up, I have to! So, let me start with what the Army has given me. It's given me an education (West Point). And you could also argue that it 'issued' my wife to me, because that's how I met her (also a WP engineer graduate). They allowed us to have and care for our kids (available medical care), to help bring them up (great programs for children), and to raise them in a safe environment. Whenever I'd show up at a new unit, they'd issue me absolutely everything that I'd need, including, in some cases, a vehicle and a driver. How cool is that! So, the Army's been the world to me. It's been awesome. It's been a great way to serve my country. And it's been a great way to provide for my family. So, yes, the Army is my life," said Aguilar, from the heart, as he concluded his remarks.

Colonel Jose Aguilar will formally retire from the United States Army at the end of Year 2020. He and his wife, Suzanne, plan to retire to the State of Texas to be closer to extended family. Their three children are all grown and living on their own in three different geographic locations. The Aguilar's have two grandchildren, to-date. Son Joey recently graduated from a university ROTC program and received his officer's commission in the U.S. Army Reserve as, like his dad, an Engineer. Their daughters have both become nurses. So, the Aguilar Family's Army connection, and dedication to service, continues.

**UPDATE:** June 2024, Colonel Jose Aguilar lives at The Woodlands (north of Houston, Texas). Applying his extensive engineering background, Jose works for STV Company, a national architectural and engineering firm, where he is the Vice President and Manager for the Greater Houston Area.

# CHAPTER 2

## From Excellent H.S. French Horn Player to Exceptional Special Forces Warrior
## United States Army
## Lieutenant Colonel Stephen R. Bolton

Lieutenant-Colonel Steve Bolton's mother and father were career musicians and music teachers, with his dad eventually becoming a bandsman with the United States Army, posted then on an unaccompanied tour to Germany, leaving his pregnant mother in the D.C. area, which explains Steve's birth at the Army's hospital at Fort Belvoir in Northern Virginia.

Army service ran at least two generations deep. On his father's side, his enlisted grandfather serving at the end of World War II, then attending, graduating, and was commissioned from, the University of Maryland, and went on to serve in both the Korean and Vietnam conflicts. On his mother's side, his other grandfather joined the Army Air Corps late in World War and, as a Lieutenant, had just completed training as a B-29 navigator destined for the Pacific, when America's two atomic bombs brought the war with Japan to a close. Steve's dad served four years on active duty, and then in the Army Reserve for many years thereafter. Parental brothers also served. His father's brother, Mike, served a full career as an Army artillery officer, while his mother's brother served five-years in the Air Force as a Korean linguist. Truly a distinguished military family.

Looking back at his high school years, Bolton recalled that he did very well academically. He especially enjoyed math, that is "until I ran into college calculus as a junior in high school and decided that I never wanted anything to do with it again!" He enjoyed sports, playing not on school teams, but recreationally. His real love and accomplishment during those H.S. years was playing the French horn, following his parents' lead and becoming "one of the top players in the state." His career interest at that time, thus shifted "from becoming an aerospace engineer to an orchestral musician," said Bolton.

In September of 1989, he began his college years at the University of Kansas, attending on a 'full ride' music scholarship. "I loved what I was doing there, was very enthusiastic about all my music classes, and never had any intention of ever joining the military," he remembers. But since he lacked the necessary enthusiasm for other classes that he didn't enjoy, by the Summer of 1990, he found himself "absent some scholarships!"

He wanted to continue with his collegiate studies, so he looked around and determined that service in the Army Reserve could provide the financial pathway he needed to remain in school. So, in August of 1990, he signed his enlistment papers and went to basic training at Fort Jackson, South Carolina, planning to be a French horn player in an Army band, conveniently located in Lawrence, Kansas, which would enable him to return to school. "I did that for a couple of years, but at the same time, I started to have some

interest in other academic fields, so I started taking several political science classes and things of that nature," Bolton remembered. Perhaps surprisingly, "A long-term military career started to look more appealing to me, because I very much enjoyed the values that were associated with military service. It wasn't just my family background. It was an entirely new appreciation for what I observed and increasingly felt."

*LTC Steve Bolton discusses the beginning of his career.*

And as his future career direction would have it, there also happened to be an Army Reserve Special Forces unit nearby. "I transferred over to that unit, because I decided If I was going to be a career soldier, I wanted the best training that I could get," he said. "I joined the Reserve Army Special Forces unit in 1993, and that prepared me for the Special Forces Assessment course, and then the subsequent Qualification course, that I attended in 1994." At that point, he had about 2 ½-years of college completed, and decided he'd take a break in order to go through about a year-and-a-half of Army training. Then, with only about one year of that training completed, in January of 1995, he got his first deployment assignment, this one to Haiti!

Looking back for a moment at his initial Special Forces training, the first stop was jump school at Fort Benning. "The Special Forces assessment course was a twenty-day, very intense, assessing process, from which only about 30% are accepted for continued training to see if they'll meet the qualifications to become a Green Beret," said Bolton. After earning his jump wings, he moved next to Fort Bragg, North Carolina, where he entered the communications sergeant course, preparing to become a radio operator.

"That time, from April to December of '94, when I graduated from the SF Qualification course, was one of the best things that I had experienced in my life. Not just the subject matter, the adventure, and the challenges associated with it, but it was the quality of the people I was around. It was just a tremendously good experience," he recalled. At that time, a romantic relationship (led to marriage, but eventual divorce) kept him in North Carolina rather than returning to complete his college studies in Kansas. He would, of necessity, deal with those missing credits later on.

Also in 2004, the Army moved Special Forces along with all of the combat arms units (infantry, armor, artillery) from the Reserve branch to the National Guard, which included Steve Bolton, since his Reserve unit went away as well. That being the case, and in keeping with his developing Army career emphasis, he then changed his original Guard unit designation to one in Special Forces.

He actually continued to enjoy his Special Forces experiences so much that, recalled Bolton: "I just kept finding more reasons to go on active duty, attend more schools, and go on more deployments." So, in 1996, as indicated, he changed his Army assignment from communications to medic, heading then next to the Special Forces Medical Sergeants Course (eleven-month duration). As he described the outcome of that training: "Being an 18D, SF medic, is kind of like being a Physician's Assistant (P.A.), with some extra expedient surgical skills included. They start us off as paramedics, and then they build from there. For me, that was also a very enjoyable, very intense period of training." And it was during a medic training rotation at Fort Belvoir, when he would meet his second, and forever wife, Brooke, who was there on staff with her first duty station as an Army nurse. Serendipitous, since Fort Belvoir, you'll recall, was his place of birth, and now, the site of a very happy 'rebirth' of sorts for Sergeant Bolton, who would go on to marry Brooke after about a year of dating.

At the conclusion of the SF medical course, he continued to enjoy all that he was doing so much so that he officially applied to transition to Army active duty. Once that change process was completed, in May 1998, he became an active-duty Army NCO and moved to Fort Bragg (NC) to join the Seventh Special Forces Group headquartered there. By that summer, he "found himself on the 'A' Team, ODA 764, and I would be with that same team for the next six-years." Wife Brooke had by then earned her qualification as an OB/GYN nurse and applied successfully (for Army permission) to practice at Womack Army Hospital, enabling she and Steve to live and work there together at Fort Bragg.

While there, from 1998 to 2004, Sergeant Bolton did six deployments with his SF team. "We started off in Paraguay in 1999 (along with Bolivia). In 2000, we did Argentina. In 2001, Colombia. In 2002, we went to El Salvador. What we were doing in all of those places was partnering with a host-nation unit and training them; basically, improving their skill set. Which is our peacetime mission, also translating very well into our wartime one. Building partner capacity. All of the trips typically lasted two to three months," recalled Bolton.

Then, in August 2002, he went to Afghanistan. Stationed on the Pakistan border, at a place called Shkin, which would become his very first combat mission (2002-2003). His team would remain in

Afghanistan, on that deployment, for about six-months. "That was the first time I had ever taken enemy fire," said Bolton. "One night, the insurgents decided they were going to attack our high-walled compound ('fort') from about 500-meters away. Firing RPG's, which was ridiculous, and very ineffective, from that distance. Looking above our high compound walls (20-30-feet high), you could see that the sky overhead was also full of tracers. It was kind of surreal. So, we're like, OK, let's get up and get our vehicles ready. We were out the door in a matter of minutes. And, again, that was my first experience, in country, with enemy fire," he recalled.

Interspersed between all of those deployment phases, there typically remained a month free at home station for Bolton and the other team members to attend desired individual training schools. That was the time when he trained for, and completed, his parachutist free-fall certification, along with several other specialty skills. During that training time, "my role as a medic was to make sure that everybody on my team, no matter what their individual job, could give each other IV's, as well as handling some basic trauma management, things like that," said Bolton. Individual and collective training, always aimed at, and getting ready for, their next deployment.

And a note on his jumps required for free-fall certification. He recalled that the standard 'HALO' jump was from about 13,000-feet above sea level, with a chute opening at about 4,000 feet above ground level. "The highest I ever jumped was from 25,000 feet (required wearing oxygen, due to the extreme altitude). The highest I ever opened my chute was 17,000-feet, which was a nice long ride under the canopy, traveling several miles. Once you do open at the more standard 4,000-feet, you're on the ground in a minute or two," he said.

When those despicable Islamic terrorists purposely flew their planes into New York City's Twin Towers, sixty-years after Pearl Harbor, making it the infamous second 'Attack on America' (9-11-2001), Bolton was at free-fall jump master school in Yuma, Arizona. That didn't immediately impact his team because it wasn't within his SF team's geographic area. "It was determined that the U.S. would take action employing the assigned geographic territory of the Fifth Special Forces Group. But we knew that within about six-months, we'd be picking up that Fall, 2002 rotation to Afghanistan," said Bolton.

During both home station time and on deployment, Bolton somehow (i.e., gritty determination) made the time to work toward completing his college degree so that he could move his career forward. "I was taking classes, essentially distance learning (!), in 2001 and 2002 while in Afghanistan. Back then, we had the floppy disks, and when one of my teammates was going home on leave or something, I would give him a disk containing all the papers I'd written. I was taking a Constitutional law class and things like that. Sending papers back (by colleague courier) whenever I could, because email was not reliable in a remote location. So, I was actually able to complete my degree in 2003, and I earned it from Campbell University, because they had a special program for medics. So, despite all the music and political science classes from before, I ended up, and proudly, with a Bachelor of Health Science as my undergrad major!" said a much relieved, remembering back those times, Sergeant Bolton. Relieved, because that four-year degree was required to fulfill his now-next-step-dream and plan. "After I came back from Afghanistan, on that first deployment there, I put in my application for Office Candidate School!"

Readily accepted, Fort Benning was the site for Bolton's OCS training, which he completed in April of 2004. "I commissioned in armor because Special Forces doesn't take lieutenants," he said, "so I couldn't go right back into that community. I had to do some other job in the Army, so I decided to go armor." He then completed the armor basic course at Fort Knox, Kentucky. From there, he took his family out to Fort Carson (Colorado Springs, CO) where he became a heavy scout platoon leader with the First Squadron, Third Armored Cavalry Regiment there. "So, I went from being this fighter SF guy who'd spent ten-years, you know, running around the jungles of South America, to now dealing with heavy armor tanks and Bradley Fighting Vehicles! But the soldiers that I had in my platoon and my troop were just fantastic. And even though that was only about a year and a half of my entire career, I still have as strong a relationship with many of those NCOs, as I do with any of my SF teammates over the other 27-years," remembered Bolton.

Serving (2004-2006) with the Third Armored Calvary Regiment (ACR) out of Fort Knox, Lieutenant Steve Bolton would spend that middle year (2005+) deployed to Iraq in that role as a heavy scout platoon leader. Based out of Baghdad International Airport, his unit's responsibility was patrolling the area to the west of the airport. "We're in armored vehicles, but there are a lot of water canals here, the roads are all on high banks, meaning there aren't places where you can get off the roads, " recalled Bolton. As a direct result, their armored-vehicle-troops were encountering IED's there on almost a daily basis. That particular one-year in Iraq was, in Bolton's word, "intense." What follows is the description of but one harrowing example of that, and sadly, a tragic one.

His team was patrolling 12-hours on/12-hours off throughout this first month in-country. He remembered that about half his platoon had participated in America's initial invasion of Iraq, and thus had some combat experience, but the other half of the platoon did not. Bolton had been in Afghanistan, and other locations, so he was a combat veteran. This "was a very high stress environment for my troops, so we continually worked on our battle drills, doing all of the things that would keep everyone's mind, and reactions, sharp," said Bolton. "We also did a lot of stuff in our off-time, when we weren't sleeping and maintaining (armor) tracks, in order to keep our resilience up, since this was our first thirty-days there, and there was already a lot of (enemy) contact. And so, it was curious to me what we were going to do to sustain that, over what was going to be a year-long deployment," he remembered.

Then came a new mission: "Operation Restoring Rights." His Third ACR squadron ("Tiger") would now go up north to a city called Sinjar, which is just about the last big city in northwestern Iraq before the Syrian border. Apart from enemy encounters, there was a significant humanitarian crisis involving the Yazidi people residing there. Unlike the area west of the Baghdad Airport, the Sinjar region is "completely different terrain; it's wide-open country," recalled Bolton. Additionally, due to the persistent enemy insurgency and the fact that the Third ACR was then headquartered near-by in Tall Afar, the majority of the residents of Sinjar had either left or were driven from the city. Their departure also hastened by the fact that most of the ACR Tiger Squadron would then be moved from Sinjar to join in the major fight in Tall Afar.

The American operation would actually require brigade-level planning and involvement. "We would seal-off the city, in order to then clear it, neighborhood by neighborhood, block by clock, and ultimately house by house," said Bolton. Before kicking off the operation, Bolton's team and others would go into the city to do some needed pre-mission familiarization patrols. The enemy had already had some success destroying U.S. armored vehicles, so the danger now faced by our troops was obvious. As Bolton recalled: "It was perhaps a low order of probability, but probability existed, and we were losing a handful of vehicles and good people." While this was to be predominantly vehicle patrols through the streets, there did exist places where they would actually have to dismount, and revert to foot patrols, requiring implementation of their repeatedly trained battle drills to protect their lives out in the open. "We had developed our tactics, techniques, and procedures for how to cordon off a neighborhood, in order to make it 'safe' for us to dismount. Then our troops would go in and do the house-to-house searching required," remembered Bolton.

He paused in his accounting of this particular combat preparation to remember and review a key element within his on-going team training procedures. "For all of my NCO's who went on to become First Sergeants and Sergeants-Major, I had trained them to become trainers themselves. I brought with me all of the Special Forces methods, making NCO's carry the full amount of responsibility that each was capable of effectively handling, to include planning missions and then leading them. I, of course, maintained overall responsibility, but with individual development came the necessary trust in their abilities to execute along with me. So then, every time we went to a new place, a new location, a new area, a new type of mission, I was the first person to plan it and I was the first person on the ground. But these trained NCOs were right there with me, ready to execute," said Bolton.

However, just a couple of days before the ACR's full-scale operation in Tall Afar was set to begin, and in fact in the very location where his platoon was due to  dismount, to then enter and clear the first block of residences, Steve Bolton and his team would all too soon discover, that when it comes to dismounted searches in enemy territory, 'safe,' is a hauntingly relative term.

It was a known fact among the American forces there that there were snipers in that area. And one of the key downsides regarding snipers is that they have the distinct advantage and, given the loner nature of sniping, there aren't enemy forces then around for the Americans to return fire. In fact, an ACR unit had just lost a soldier to sniper fire several days earlier! Tragically, it was about to happen again.

Lieutenant Steve Bolton was riding in a patrolling tank that morning with his good friend, Lieutenant Charlie Rubado. Even though Bolton was senior to Rubado by a few months, on that particular patrol, in that tank, Rubado was serving as commander, while Bolton rode in the nearby turret as loader. Both had a machine gun mounted near the rim of their respective hatches. The two lieutenants and crew were, as mentioned, on a familiarization patrol, scoping out their upcoming areas of responsibility, prior to the start of the full-scale area-clearing operation. At one point, Robado came to realize that their Abrams tank had gotten hung up on something, and he was going to need to back it up to get clear of the obstruction. As Bolton sat near-by with his hatch also opened, Charlie Robado poked his head up to see what the issue below might be.

Seemingly, in that instant, even over the noise of their tank idling, the sound of the sniper's rifle shot could be heard. Its bullet arriving at their tank turret in an instant. It flew perilously close to where Bolton was perched in his loader's hatch. The shot narrowly missed him, as he was peering forward at that moment. Still not sufficiently exposed, he wasn't the enemy's target. Bolton's tank commander and good friend, Charlie Rubado, was.

Unknowingly, Rubado had picked the wrong moment to stick his head fully up to look around and assess their situation below. At that instant, as Bolton still vividly remembers, the crack of that sniper's rifle sent a bullet which would hit and grievously wound Charlie Rubado. As he slumped down through the hatch, his knees landing on the back of the tank driver below, without hesitation, Bolton, a combat medic in his enlisted days, sprang to assist his comrade.

"I grabbed Charlie and pulled him down the rest of the way inside.

He's lying across me and I've got his head in my lap. I see there's blood on the back of his helmet (exit wound). I'm applying pressure there, but I can't immediately find where the entry wound is," remembered Bolton. He's livid with himself, at the time, since the well-stocked Special Forces medic bag he always kept close by was then, when he needed it the most, sitting back up in a rack <u>outside</u> the turret! Meanwhile, the assembled group's troop commander had just announced the conclusion of their three-tank reconnaissance mission, and it was time for them all to pull away from the area and return to base.

Holding Charlie, but knowing he also needed to immediately get his tank and crew to safety, he told the gunner to be their eyes, to work with the driver, and get them withdrawing along with the other unit vehicles. He then called the troop commander on the radio, requested they move to an alternate frequency, one which wasn't openly monitored. He explained their situation and requested that the commander help get his tank to the designated evacuation zone as quickly as possible, so that Charlie could be flown out by helicopter to the nearest advanced emergency care facility. "All the while, I'm not putting my head up to guide us (turned that over to the gunner & driver), because I'm trying to save Charlie and keep him alive. I'm just trying to get control of the bleeding."

Bolton's tank arrived back within the U.S. zone, a definite boost to the crew's safety and morale. Medics there were already waiting. They carefully lifted Rubado out of the tank, as the evacuation helicopters were arriving. He seemed not to be breathing while in the tank but began again when stretched out on the ground alongside the medical personnel. Bolton realized that they had to "protect his airway, so he could continue to breathe," while airlifted to the medical facility. "There wasn't an endotracheal tube immediately available, so I had one of the medics cut off the drinking tube from a camelback, and then I digitally intubated Charlie to sustain his breathing. We had secured his airway, controlled the bleeding, and got him on the evac helicopter, which was all we could do for him at that point," recalled Bolton.

With his survival prognosis in doubt, despite everyone's best efforts, principally those of Steve Bolton, regrettably for all involved, and for

Bolton especially, Charlie Rubado died in the helicopter on the way to the field hospital. "I was pissed that I couldn't have done more to save him. Pissed because he was just such a great guy; just an absolutely great, great friend," said Bolton, with fond memories, to this day, of their service together.

Normally, Bolton and Rubado would not have been in the same tank. It just happened because they were conducting that special recon mission together, looking closely at the area that a much larger unit would return to the following week to carry out that full-scale clearing operation. In fact, they would be going in with hundreds of soldiers and thousands of vehicles, fully expecting that they would be facing a large insurgent fight. Those huge neighborhoods, now vacated by the former resident families, were believed to be sites where the insurgents were lying in wait. "That was their safe haven, and we were going to go in and clear out their assumed safety zone and push them back out of town. Since, as team leader, I would be the first man off the ramp (Bradley Fighting Vehicle), out in the open for this entire operation, I knew I would be their first target. So, no question, that first day clearly stands out as the fastest I've ever run wearing 80 pounds of body armor and full kit! And my team was right there on my heels as we began to hit those houses," said Bolton.

"Those were sixteen-hour days. And the daytime temps ran about 110-115 degrees!" he remembered, still feeling mighty fortunate that they encountered no real enemy resistance, aside from an occasional sniper with poor aim! Block by block, they cleared all of the houses, picking the best one to secure each evening, then kept watch, while trying to get some sleep, and got up and did it again. Did it again, that is, for seven days straight. The operation was finally, successfully completed, and, thankfully, that feared big insurgent fight never materialized. (NOTE: Steve Bolton was attached to the Third Armored Cavalry Regiment during this particular combat action, including the loss of his close friend, permitting full disclosure and discussion. As you might expect, by long established policy, Special Forces missions and encounters may neither be specifically identified nor discussed).

When Bolton returned to Fort Carson, after that "intense" year in Iraq, he knew that he would then be eligible to rejoin a Special Forces unit soon after. Fortunately, there was an SF Group at Fort Carson. He was anxious to begin "re-familiarizing" himself, and "re-integrating," as he said, within the SF community. Joining the SF Tenth Group there, in the summer of 2006, given his time as an armor officer, they made Bolton the XO of one of their SF companies. Then in 2007, he and his teammates were off, yet again, to Iraq. Stationed there near the City of Mosul, they would remain in the fight for about the next eight months.

Promoted to Captain during that 2007 Iraq deployment meant that it was time for him to attend and complete the Army's Captain's Course, which took him back to Fort Benning, this time for five-months of study, and then it was back to Fort Bragg, where in August 2008, he once again entered Special Forces Q Course qualification. SF re-qualification? Yes, because as an officer now, versus his first time through, he had to prove his proficiency with a "different skill set." That re-qual course successfully completed, by the Summer of 2009, "it was time to figure out where they were going to put me because of my origin with Seventh Group," said Bolton. 'Because I had spent a year and a half with the Third Battalion Tenth Special Forces Group at (Fort) Carson, Tenth Group was able to pull me back, and I was fine with that. I loved both organizations. So, I'm back at Carson, now as a detachment commander."

Bolton recalled that "our mission during that time was to Africa, so this was my first time deployed to that continent. And what we were doing there was a lot of the same kind of training of local forces that I had done previously in South America, but also doing, as well, a lot of embassy liaison work. Interagency collaboration in Sub-Saharan Africa, where al Qaeda had a lot of destabilizing activities."

For the first half of 2010, he was in Mali, but working, this time, in very small groups, just a few of them there, with other parts of his team in Mauritania. "This was a bit of a unique mission set for us. I only took half my team, so that when we came back in the fall of 2010, the other half went over to replace us. We had to stretch out this mission for a certain amount of time while another unit was getting ready to pick it up," remembered Bolton.

The next big mission for Captain Bolton didn't happen until the Fall of 2011, when his Third Battalion got the assignment to go to the Horn of Africa to Djibouti. Ironically, however, although involved with this specialized mission to Africa, his battalion was actually a mountain warfare team!! So, said Bolton, "we still have a whole host of tactical skills to keep up with. It's not just the normal, you know, shooting, medical, demolition, and marksmanship stuff. But on top of that, we have the responsibility to maintain mountaineering skills. So, if we're not deployed, then the team is going through a progressive training program to make sure that we're still ready to deploy anywhere else we may have to," he said.

With his company commander then deployed elsewhere, Bolton stepped up and served as the acting company commander for about six-months. From there, in 2011, Bolton moved up to the battalion to become the Assistant Operations Officer. And during the first half of 2012, his battalion staff, minus any other tactical units, heads, again, to Djibouti, Africa to become a Special Operations Task Force for East Africa. "We now have the responsibility for planning and resourcing all of the individual country training missions that our teams go on. We're supporting that across the entire East Africa region, including all of the other special operations activities. And that's not just the Army Special Forces stuff. We're responsible for the Navy Seals and the Air Force activities as well," recalled Bolton. As he thought back about those times, "this was my first foray into the operational level, the regional level of war or operations."

He returned home to continue serving as Assistant Operations Officer until the Summer of 2013. During that year, Tenth Group's area of operational responsibility was changed back to Europe. "So now we are re-educating our force on everything relevant for us about the European environment. We are reconnecting with the special operations elements that already exist in the area, but now with new problem sets. We are also starting to prepare ourselves for the possibility of Arctic warfare. So, this whole year was really about transitioning us out of an Afghanization and Africa mindset and preparing ourselves to re-connect and cooperate with European forces….and the inevitable winter warfare environment!" remembered Bolton.

Having helped his battalion commander to put the team on that path, the Summer of 2013 brought two important, non-deployment opportunities. First was Bolton's promotion to Major! And second, he had also earned the opportunity to, once again, step off the deployment treadmill and head for some additional schooling. This time it would be the Command & General Staff College at Fort Leavenworth, Kansas. And while there, never letting an advancement opportunity slip by, Major Bolton would also be pursuing a tight one-year master's degree at the University of Kansas, this one in Global and International Studies. Then, he earned selection to the "Sam's" program of study (School of Advanced Military Studies), also at Leavenworth, meaning he could keep self and family in place for one additional year, enabling him, then, to earn <u>two</u> master's Degrees in just <u>22-months</u>! Wife Brooke brings that concentrated grind up every time her husband suggests studying for a Ph.D. ("Are you really sure you want to do all of that again?"). Recalled Bolton: "Those were two incredibly rewarding and enriching years for me. I really enjoy the opportunity to get back into studies that make you think about professional matters." And he did come out of that intense academic exposure "thinking differently," when, in 2015, he went back to the 10th Special Forces Group at Fort Carson, this time taking command of the Group's Alpha Company, Third Battalion.

And within a week of his return, and taking command, he was off to Africa, this time to Chad. "I led a small Special Operations Task Force that was responsible for as variety of missions, both there and in neighboring countries (Nigeria). I had SF Teams, Navy SEALS, psychological operations, and civil affairs units, all under the Task Force. I also had liaison with embassies, as well as with something called the Multinational Joint Task Force. And the reason we were there was all about the infamous Boko Haram activities in Nigeria, and how that instability was threatening to spill over into Chad, Cameroon, and Niger," remembered Bolton. It was during this era that the initial discovery, spread, and eventual crisis from the COVID virus first became known there, and began impacting operations, resulting in reduced manpower, and other procedures aimed at preventative medicine.

In 2016, Bolton would return to Fort Carson to begin, in earnest, his

SF Group's transition from Africa-centric assignments to future work in Europe. He was now the battalion's Operations Officer, taking team members to perhaps a dozen countries, to include training & practicing winter warfare skills, and assigning forces to upcoming missions. Then, in 2017, it was back to Fort Carson, this time with the Fourth Infantry Division as its special operations advisor to the commander and staff, as readiness training continued, to include preparation for a division rotation to Afghanistan.

Beginning In 2018, and for the next two years, Major Bolton was assigned to NATO Special Operations Headquarters in Brussels, Belgium, as the Aide-du-Camp to a three-star Vice-Admiral. The mission then was to re-orient thinking and training aimed at potential threats from Russia. His role required considerable travel, forming relationships with allied military personnel throughout neighboring nations. His second year at headquarters had him steeped in plans and policies, during which time he became the lead author on a special operations study regarding the most effective way(s) for NATO countries to deploy their Special Forces should the Russian situation become active and threatening (which of course it would, and did, within four-years!). By 2020, work was slowed measurably, there and virtually everywhere, as COVID lockdowns and other precautions took control of NATO HQ. Belgium was in lock-down mode during much of that time, with outside movement permitted only for essentials.

It was during that period of health peril, both in Belgium and around the world, that Major Bolton was competing for a slot on the Command Select List (CSL), which would enable him to hopefully take a battalion command back in the states. That wish and desire would come true, when, along with his selection for promotion to Lieutenant-Colonel (July 2020), he was informed (April/May) that he had been selected to lead the Garrison at Hunter Army Airfield in Savannah, Georgia, which was supposed to occur in the Summer of 2021. But serious family medical circumstances impacting the existing Hunter Garrison Commander's wife (a LTC at Fort Stewart, GA) made it necessary, with very little advance notice, for LTC Bolton to pack up and head solo to Savannah, well ahead of the originally anticipated time, compelling him to leave his Army family back in Belgium for the next six months. They would finally be able to rejoin him in December of 2020.  So, for Bolton: "In July, I got promoted.  Then I got my orders to Savannah.  And I got on a plane, all in the space of about 13-days, with next stop Hunter Army Airfield, USA!"  A blur of Army intercontinental transfer forms & procedures, and of course, buttoning up his family to keep them secure in Belgium for the next six-months, filled every waking moment of LTC Bolton's last two-weeks at NATO.

Hunter Army Airfield Garrison (Savannah, Georgia): "As long as I'd been in the Army, I had no idea what a Garrison was, and that's because it's not something that most soldiers are assigned to, and certainly not Special Forces soldiers!  Well, come to find out, a fair number of SF guys had actually commanded the HAAF Garrison, in two-year assignments, before me!" said Bolton.  Given the uniqueness of this command, he set out to make certain that his largely civilian Garrison team fully understood their role, and their impact on, and with, the multi-branch tenant units stationed there on the installation.  Chain of command wise, the Hunter commander answers to the Garrison commander (Colonel) down at Fort Stewart (Hinesville, Georgia), home station of the Army's Third Infantry Division.

Bolton made it his focus to further educate, not only those serving around him, but importantly, those out in the community who may have little knowledge about the Garrison role at Hunter, and Hunter's Army role as home station to  aviation-based and other military units serving there, active-duty, guard & reserve.  Regarding the other tenant units and personnel on Hunter, along with their families, the commander's role is much like that of a town mayor, dealing with infrastructure issues and needs, as well as medical, educational, recreational services, all wrapped up in the overall quality of life on Post needed and deserved by the military members, their families, and the civilian employees, all of whom  serve the Army and our nation.

LTC Bolton especially appreciated the peer relationships he formed working alongside his colleague unit commanders, especially "since we all not only work together, but also socialize together, living as we do, side by side, within the same Post neighborhood. I find it very fulfilling to serve in this command capacity dealing with both the internal on-Post community, as well as the Greater Savannah community outside the front gate," he said. He has made a genuine outreach effort to civilian leadership and groups, eagerly inviting them to tour Hunter, since so many really have no direct knowledge about the units and missions that call Hunter home. And the more they know, the more likely they are to support Hunter when needs or issues, or even possibly emergencies, arise.

Reflecting on his time in Garrison command, as his tour at Hunter draws to a close (July 2022), Bolton was quick to value his outreach efforts: "I have enjoyed the interaction with the Savannah community immensely. It's really been a privilege to be part of both the Chamber of Commerce and the Downtown Rotary Club, and to have the opportunity to interact with civic leadership and with several civic organizations, because it's helped me understand how communities work, and, importantly, how those who want to influence communities for the better can bring their collective vision to bear. And most of all, I think I've enjoyed the opportunity for my family to experience Savannah, along with the always welcome opportunity to more broadly share the Hunter story, as well as also telling, and representing, the broader Army story," said Bolton. And with regard to the personal value and pleasure he's felt being stationed here: "Throughout my three decades of military service, nowhere else have I seen a city that cares so much for its Army community," he concluded, with sincere appreciation.

LTC Bolton's next assignment will take him back to Fort Leavenworth, Kansas, where he'll be the Director of Special Operations Education for the Command & General Staff College there. "My actual official title is Executive Officer of the Special Operations Cell at the Combined Arms Center. But that doesn't really mean anything to most people who aren't in the organization. More to the point, I'll be kind of a curriculum manager. I'll be making sure that we're teaching the right things to the right audiences," said Bolton. Most of the "right audience" students that he will be dealing with, and instructing, will be majors, junior majors, and senior captains, all of whom will be at about the 10-year mark in their respective careers. "The purpose of this course of study is to broaden their understanding beyond their own branch, their own specialized knowledge, to give them enough cross-training awareness so that they can then become field grade officers (Major, Lieutenant-Colonel, Colonel). Essentially, these are the soldiers that will go to work for General Officers or serve on General Staffs (at the Division or Corps levels)," he said.

To round out this career discussion session, LTC Bolton was asked a couple more general questions regarding leadership, beginning with his thoughts on the key elements or qualities that make leaders most effective. Without hesitation, he replied: "Trust, empathy, and compassion (but not 'softness', he emphasized). Taking care of your people, and communication (both with clear delivery, and openness to response)." Then, what did he feel are the main components of solid character, so very important for effective leadership? Again, without any hesitation, he listed: "Integrity, respect, perseverance, and moral courage."

Finally, given the qualities and the character components required for effective leadership, LTC Bolton was asked to think back to perhaps a stand-out example of such that he had personally observed in an officer throughout his own stellar three-decade Army career. The name that came immediately to mind was LTC Greg Riley. Bolton remembered him specifically for an act of great leadership and courage that he had personally witnessed, performed during an Iraq War deployment.

One day, in order to get a better indication of where insurgent snipers might be hidden, two Iraqi soldiers decided to climb up a tall utility tower to get a better view. And, yes, they did get a better view. But so did the sniper(s)! Those two Iraqis had mistakenly become convenient targets. Seeing the issue and the rapidly developing emergency, on the ground, LTC Riley tried in vain to get Iraqi commanders to get their troops to rescue their own soldiers, but all refused! Thinking innovatively, Riley thought immediately of a chemical detachment located very nearby. One of that unit's many combat responsibilities was to produce defensive smokescreens. He quickly summoned them. Their subsequent smokescreens resulted, as planned, in obscuring that tall tower from enemy view. Then, Squadron Commander LTC Greg Riley, himself, heroically climbed up that tower to rescue those two Iraqi soldiers, one or both of whom, by that time, had been wounded by enemy fire. It was an instinctive act of great courage, recalling that those two soldiers, although allies, weren't his! There is an old, yet still very applicable and profound, saying, that applies to both military and civilian life. It relates that, when faced with a needed decision or action, we will hopefully "choose the harder right rather than the easier wrong." In this vivid instance of personal courage, LTC Riley clearly showed all those around him, including the Iraqi commanders, exactly what choosing the 'harder right' looks like!

Lieutenant-Colonel Steve Bolton's Change of Command ceremony at Hunter Army Airfield took place on July 29, 2022. He and his family will then head directly for his next assignment at Fort Leavenworth, Kansas.

*LTC Steve Bolton*
*US Army Special Forces*

**UPDATE:** LTC Steve Bolton has completed his assignment as U.S. Special Forces Director, Special Operation Element, at the Army's Command and General Staff College. He moves now to Carlisle, Pennsylvania having been selected to attend the Army War College there for the 2024 – 2025 academic year.

# CHAPTER 3

## Help From Above, Whenever the Need: Always Ready! Captain J. Marshall Branch, Rescue Helicopter Pilot United States Coast Guard, (Ret.)

Marshall Branch knew from an early age that he wanted to fly, and, most importantly, he wanted to do so in America's military. Military was always an important part of his family growing up, from his grandfather serving in World War II, flying with the Mighty Eighth Air Force, to his father who served as an Army officer during Vietnam. And it was his dad who "instilled an incredible sense of civic responsibility in our family. We just grew up in a family of service," recalled Branch. For him, flying as a career was never a question of if, but always a matter of when, where, and for which of the services. Early on, he had, perhaps naturally, gravitated toward the Air Force, since his dream then was to fly fighter aircraft. "The movie 'Top Gun' had come out around that time, and of course every teenage boy then wanted to be a fighter pilot," he said.

*Young Marshal Branch with his dad, while living and serving with the U.S. Army in Germany*

Then fate intervened, as it often does, this time in a very positive way for his future. With the family now living for a couple of years in New Jersey, like his dad's earlier achievement, now an Eagle Scout in the making, Branch's Boy Scout troop had a trip planned to Mystic Sea Port in Connecticut. Following that experience, on the drive back to New Jersey, they noticed an exit sign for the United States Coast Guard Academy. Not on a tight return schedule, his dad decided to take the troop over for a look. They toured the campus, and on that day, a Coast Guard Cutter was tied up at the pier, giving them an opportunity to talk with the crew. "I started to realize that there was a lot about the Coast Guard that I really liked, with missions such as law enforcement and search & rescue. That visit put the Coast Guard on the radar for me, and certainly the Academy as an option," recalled Branch. Then just a seventh grader, his dad reminds him that, while departing that day through the Academy gate, he turned to him and said: "Well, Dad, that's where I'm going to go to college!"

Already exhibiting a serious commitment to public service, with his work as a volunteer firefighter while still in high school, during his senior year, as predicted five-years earlier, Branch applied to the Coast Guard Academy and was accepted. That alone was a distinction. Each year, roughly 10,000 young men and women apply for admission, with acceptance based on key criteria like academics, athletics, civic responsibility, and leadership, all areas in which he had excelled. From among those ten-thousand applicants, only about 300 are invited to enter. Based on his accomplishments, in the obviously very competitive selection, Branch gained admission to the Academy's Class of 1991.

*Marshall Branch's Grandfather,*
*Army 1LT Sam Marshall*
*(WW II B-17 Bomber/Navigator, Mighty 8th Air Force)*
*pins on a shoulder board*
*(Branch's Dad pinned on the other),*
*at Marshall Branch's Academy Graduation (1995).*

His fondest Academy memories centered on summer training assignments out with the fleet, out with actual Coast Guard units. Those experiences included service on the USCG patrol boat that, as the maritime security command post, was tasked with watching over President George H.W. Bush wherever he went, while vacationing in Kennebunkport, Maine. The time spent with a Tactical Law Enforcement Team (TACLET) in Miami, chasing drug runners in cigarette boats. And part of a summer spent on the Coast Guard Cutter VIGILANT out of Cape Canaveral, FL, and the remainder with H-65 crews at Air Station New Orleans, the latter a welcome preview of his career to come. After four years of study and hard work, Marshall Branch graduated from the Academy in 1995 with a bachelor's degree in management and a commission as an Ensign in the United States Coast Guard.

As with all Academy graduates, his first tour was two years at sea aboard the Coast Guard Cutter DAUNTLESS. Among their many sea-borne missions, for decades, now, and still to this day, the Coast Guard has been heavily involved with drug and migrant interdiction efforts. Lieutenant Branch recalled one operation in particular, from those very early days in his career. This was a migrant interdiction.

"You have two different types that you have to worry about on migrant boats. You have the desperate people who gave up everything they had to buy passage on that vessel. They've got nothing to go back to. And then you have the 'coyotes.' They are the enforcers. They get paid to not let people jump ship. They'll threaten the migrants in all kinds of ways, even pouring gasoline on them! The only time I ever drew my weapon, as a shipboard law enforcement officer, was during migrant interdictions off the north coast of Haiti, during my very first Coast Guard tour."

"I was in one of our small boats," continued Branch. "We'd have two Coast Guard small boats on either side of the migrant vessel. We'd bring one in, making a lot of noise, and the 'coyotes' would come over swinging their machetes. Meanwhile, our other boat would sneak up on the back side and pull migrants into the boat to get them to safety. And then we'd switch sides. We'd have the 'coyotes' running back and forth constantly. Well, this one 'coyote' figured he would hide within the migrants, and when I pulled the boat in, I was reaching for one of the migrants, a young girl, and the 'coyote' popped up with his machete and was about to swing it down on that young girl. I immediately drew my sidearm. Fortunately, he backed down."

"That group, and others we encountered, had killed migrants before. Anyone disobedient or who made the 'coyotes' mad, they would just throw them overboard. There were always sharks that followed migrant vessels! So, then, particularly during the mid-1990's, it was bad. It was very dangerous out there."

Branch remembered that the largest single interdiction in Coast Guard history, and that record still stands, was an 80-foot vessel with about 500 migrants onboard. Said Branch: "They were packed in so tightly, that the ones who died just stayed there in their spot because no one could move. We didn't know how many dead people were on board until we got the rest of them off. It was definitely a desperate time, so we tried to get to those vessels as quickly as possible, because migrant safety was definitely at risk. There were a lot of boats that did not make it," he recalled.

Jumping well ahead in his career, for a moment, since we're on the subject of interdiction missions, Branch was involved in counter-drug efforts as well, both on the water and flying, as he explains.

"I remember chasing after them (drug boats) in flight, back in the days before we had armed helicopters. Now, the Coast Guard has an armed helicopter squadron in Jacksonville. They'll put a 50-caliber round through the engine block of a drug-runner's boat. That stops them! Prior to that capability, we would just basically 'squat' on them and use our rotor wash to try to disrupt their ability to maneuver, giving our boats enough time to catch up," recalled Branch.

But what if someone on those fast drug boats you were flying down close to was armed? "Back then, there was a gentlemen's agreement, as strange as that may sound. If all they're doing was evading law enforcement, they'll be arrested, but more than likely, they'd be home before our boats got back to home station! However, if you fire on a federal law enforcement officer, you're not going home, but going, instead, to the penitentiary for a great many years," he said. "Fortunately, I never had a shot fired at me from a narcotics runner," said a still understandably relieved Branch.

Back, now, to the proper chronology in Lieutenant Branch's career development sequence. Following his sea duty, in line with his long-held dream, he was off to fight training at Naval Air Station

Pensacola (FL), where he learned to first fly fixed wing aircraft, followed by rotary wing training, and was awarded his aviator wings in 1998.

Following flight school, Lieutenant (jg) Branch was assigned to Air Station Port Angeles, Washington, where he flew the H-65 helicopter and, by merit, gained the designation as an Aircraft Commander. Those stationed there refer to the area as "Alaska Lite," due to the similarities with Alaska weather and terrain, except they were close to Seattle and "civilization!"

His first rescue flight occurred at Port Angeles. Not long out of flight school, Branch was the co-pilot on what was to be just a training flight, during which he would take the pilot's seat for practice. Once in the air, they got a call from base that a man, woman and baby in a small boat had run aground after hitting some rocks, a jolt that propelled the child out of its carrier seat, with resulting concern about a possible head injury.

So, the radioed instruction was for the aircraft to hoist the mother and child off the boat and transport them to a hospital in Seattle. "I had never hoisted a live person before, in fact I had just become qualified to be in the right seat (pilot), so I was very nervous that my very first human hoist was to be a baby with a possible head injury," said Branch. The aircraft command pilot, now in the co-pilot seat, asked Lieutenant Branch if he was OK piloting for the hoist. "I told him, yes, although my heart was pounding about a thousand miles an hour! But the training kicked in, we got over-head above the boat, and I remember looking down and seeing the look in the mom's eyes, and thinking, yeah, we've got to do this," he still remembered vividly, now almost two-decades later.

His crewmates skillfully hoisted the mother and baby up into their helicopter. Branch recalled that "the experience of looking over my shoulder as the basket comes into the door, seeing the survivor(s) in the basket, and you see that look of relief, just sheer relief that things are going to be OK. That hooked me for life. That feeling of knowing that you made a huge difference in somebody's life, just guaranteed that I'll be doing this for as long as they'll let me."

It was an amazing day in the operational life of this still relatively new Coast Guard pilot. And the great news later to punctuate Lieutenant Branch's first rescue? After flying them to a Seattle hospital, they learned that the baby had survived the head injuries! They learned that, since the nurses meeting the helicopter on the ground had realized how concerned Branch's crew was about the baby, they made sure word got to them. The father remained with the boat and was towed in to shore safely.

On Lieutenant Branch's first duty night at Port Angeles, now as an Aircraft Commander, and the officer in charge, the Navy called with a rescue case, but would indicate nothing more than that the helicopter needed to be at a specific latitude and longitude at a specific time (sunrise), about 30-miles out in the open-water of the Pacific Ocean. "We launched before dawn, arrived at the latitude and longitude, and literally, the moment the clock ticked to the specified time, the USS OHIO surfaced right below us!" remembered Branch. His memory still vivid with the image of that huge American ballistic missile submarine, with its massive steel hull, that had so suddenly fully revealed itself beneath him.

*McMurdo Sound in Antarctica.*
*Lieutenant Branch visits with "friends" near the icebreaker,*
*Coast Guard Cutter POLAR SEA (2002).*

From radio communication with the sub, as Branch maneuvered into position for a hoist, he learned that one of the boat's crew had fallen down two levels of steel ladders within, hitting his head on the first landing and kept tumbling down the second. Once he was brought up top and was positioned for the lift, and the rescue basket was lowered down, "we immediately hoisted the badly injured sailor, and literally, as the basket left the sub's deck, they were already breaking things down, and by the time we got the sailor to our door, they had the hatch closed and were already starting to go below the surface again," recalled Branch. That sailor was then flown to Madigan Army Hospital on shore, with no subsequent report on his eventual outcome.

Then, in 2002, came a totally different experience and climate(!) when he was transferred to the Polar Operations Division at Aviation Training Center Mobile (AL) for a one-year deployment with his aircraft and crew abroad the USCGC POLAR SEA for a tour in Antarctica. There, at McMurdo Station, Lieutenant Branch and crew were positioned on land while their host ship went about the necessary work of cutting through the continuingly forming ice accumulation in order to keep a channel open so that supply vessels could always reach McMurdo.

*Cape Crozier on Ross Island, Antarctica.*
*Lieutenant Branch & crew flew scientists out there to study*
*the Adelie penguin colonies. Spent the day helping out, before*
*flying the team back to McMurdo Station (2002).*

While on watch there, Lieutenant Branch received a radio call indicating that a contract firm's civilian helicopter had gone down, some 50-miles distant from their McMurdo location, across the Ross Sea. With 24-hours of daylight in Antarctica, this would be a "daylight" rescue launch, and one into less-than-ideal flying conditions, punctuated by heavy fog, some wind, and light snow falling. By policy, civilian contractor aircraft were not permitted to fly in the weather conditions that Branch, and his crew would face, making the Coast Guard, then, the only rescue source.

For navigation, in an era back before today's satellite-assisted technology, they had to rely on the old stand-by, charts, because compasses don't work all that well ("the variance is 178-degrees up," recalled Branch). "The majority of the navigation in Antarctica is visual, with chart back-up," he said, making a mandatory rescue flight such as that one inherently risky. Apart from weather conditions, here's why: "The most challenging thing about flying in Antarctica is that the ice forms just a flat, white layer beneath you when it meets the shoreline, and then blends seamlessly into the glacier, creating what appears to be a continual white slope, so there's not much contrast or relief, and when your visuals are limited by weather, it makes it extremely hard to navigate safely." That continual slope-appearing ice can also distort the old-style altimeter readings, because their radar beam can actually pass right through the ice, rather than provide a bounce-back reading, making altitude even more difficult to judge and maintain.

Approaching the contractor helicopter crash site, Branch could tell, visually, that the aircraft had sheered its engine mounts, indicating that all power had been lost and, from a high hover, which Branch later estimated to have likely been somewhere around 175-to-200 feet in the air, the helicopter had dropped straight down for a very hard land impact. "The pilot (sole occupant) was trapped in the wreckage, with compound fractures and, needless to say, he was in bad shape, really bad shape," recalled Branch. Assuming there would be injuries, two McMurdo Sound-based firefighter-paramedics flew along on this rescue mission. And good thing, because four of the five on board, including Co-Pilot Branch (the Aircraft Commander stayed onboard to keep their helicopter running), had to work hard and fast to free the injured pilot. "We had to use crash axes, and whatever else we had available, to try to start pulling the downed aircraft apart just to get him out," said Branch.

*Site of the helicopter crash in Antarctica.*
*Front canopy section forcefully removed by Branch and his supplemented crew,*
*in order to rescue the badly injured pilot. In the background, a barely visible rock*
*cliff, shaded by fog, that Branch had to avoid en-route to the downed helicopter.*

And once they were able to do that, he was in for a surprise. Come to find out that the badly injured contractor pilot was retired from the Coast Guard!

"Long story short, we ended up tearing the heck out of that helicopter, we got him out, got him on a backboard, put him in our helicopter, and began the flight back to McMurdo for medical care. We hadn't realized until then that the injured man was a retired Coast Guard pilot. He'd been passed out from the pain and loss of blood, until he heard the familiar sound of the approaching Coast Guard engines, and as we were loading him into our helicopter, he came to, his eyes briefly darting around, recognized he was in the same type of Coast Guard helicopter he had flown, gave us the thumbs up, and passed out again!," Branch clearly remembered.

The McMurdo paramedics onboard worked to stabilize the injured pilot. While the rescue, onboard medical attention, and return flight to McMurdo were all underway, a C-141 "Starlifter" was launched from Christ Church, New Zealand, with medical personnel, including a trauma surgeon, and advanced treatment facilities, effectively a hospital in the sky, which would then transport the injured pilot from Antarctica back to New Zealand for continued, land-based, medical care.

But how and where to safely land an aircraft of that size? Branch explained. "McMurdo Station has the world's only ice runway. Pegasus Airfield is actually out on the Ross Sea. The ice is about 20-feet thick, so they can actually land and launch cargo jets on it. By the time we landed at McMurdo, beating the C-141 there by just a little bit, we were set to make an immediate aircraft-to-aircraft transfer. Once he was onboard that large jet, he was in very good, experienced hands." Best news of all for everyone involved in that risky rescue, Branch later learned that the injured pilot did pull through.

*Lieutenant-Commander Branch (left seat in view) en route to a mission (2008).*

Following the Antarctica assignment, in 2003, LT Branch was selected to enter the aviation engineer career track, along with his flying duties, and was transferred to Air Station Traverse City (Michigan). He wanted the aviation engineer option, rather than operations officer, because it would mean spending most of his career "on the hangar deck." This kept him constantly around the helicopters, flight mechanics, and crew. Being around all of those folks for the bulk of his career was exactly where he wanted to be. It was at the Traverse City Air Station that, now-Lieutenant Commander Branch became an Aircraft Commander, Instructor Pilot, and Flight Examiner in the upgraded HH-65B.

Northern Michigan has some really challenging winter weather
conditions. Despite that, outdoor activities like snowmobiling, ice
fishing, cross-country skiing, hunting and other recreational hobbies
are popular, and on occasion, can result in problems, especially with
sudden changes in the weather.

One night, while LCDR Branch was on duty, he got a rescue call
regarding a hunter who had failed to return home. "He was out on
one of the Western Michigan islands (reached by personal boat). It
was snowing pretty heavily that night, so we launched out, because
they couldn't get a boat to the island. The storm had blown up so
much that there was no way he'd be able to use his own boat, with
the approximate 5-to-8-foot seas. Making matters more urgent, with
temperatures approaching zero, and gusty winds, they learned that
the hunter wasn't equipped for overnight survival. Fortunately, there
was a State of Michigan wildlife officer housed on that particular
island. The challenge, now, was getting that officer together with the
hunter in the midst of what had become a blinding snowstorm.

Thus, the pressing need for Branch's crew to get overhead to try to
locate that hunter. "The snow was so heavy, making visibility very
limited, and it was at night, so we were flying with night vision
goggles. And flying through snow like that almost completely
overwhelmed the vision part of the term night vision goggles,"
remembered Branch. Beyond the snow, adding to the difficulty of
spotting the guy was the fact that the island was very heavily wooded.

It all came down, luckily, to one habit on that hunter's part that may
well have saved his life. "We finally found him, when we saw him
light a cigarette! And on the night vision goggles, that popped pretty
clearly on a very dark island. Then, through some radio relays, we
were able to direct the wildlife officer over to him, so that he could
get him back to the officer's hut to warm up. We did learn that he
was pretty hypothermic. If we hadn't found him that night, he
wouldn't have made it," concluded Branch. Ironic, then, that, on this
totally dark and freezing night, in a twist to the norm, this was a life
actually saved, thanks to smoking!!

Since they had flown all the way out there, in very challenging winter
conditions, for their effort, had they considered simply hoisting the
hunter up and off the island? The answer to that question was a

pretty quick No. "With every rescue situation, you have to make a judgment call. Hoisting is dangerous. People think that just because, in this case, you found him, you need to hoist him. Sometimes, the boat transfer, the shore transfer, or the shore forces taking that person into their care, is better because its less stress on someone already impacted by their situation, or even in shock, so that sometimes flying them out is not the best plan," said Branch.

Even as an Air Station Commander, that person remains in the rotation to "stand duty." One night, in January of 2018, while Captain Branch and his crew were on-duty in Charleston (Air Station Savannah rotates a crew every day to staff and cover Air Facility Charleston), a call came in from a fishing vessel that was about 75-miles off of Cape Fear, North Carolina. "It's the middle of the night, with 10-to-15-foot seas, and probably about 45-mph winds. We were told that the captain had gotten his arm caught in the reel mechanism that helps to bring up the long lines. He'd mangled it pretty badly and was bleeding heavily. We consulted with the on-duty fight surgeon (available by phone 24/7 to evaluate requests for medical evacuations) and he said, yeah, you need to get this guy off," said Branch. Because of the distance from shore, and with it, the lack of helicopter-to-base radio reach, joining the rescue team was a C-130 from Elizabeth City, North Carolina, to fly overhead providing radio relay back to the coast. "Having fixed-wing coverage when we're that far offshore is really important for the safety of my crew. They can also help prep the 'battlefield', because one of the rules of SAR (search and rescue), that we grew up learning, is that most of the things you're told about the case, en route, aren't true, and, unfortunately, there's some truth to that old adage!" said Branch.

And case in point, when they arrived on scene, the vessel was not at all what they were expecting. Based on past such missions, they thought they'd be finding more of a regular commercial fishing boat, with a big open deck in the back, and trawler arms. Instead, they were looking down at a vessel with a covered deck from the very front of the boat all the way to the back. For starters, it was then obvious that there was no way for the boat's injured captain to get up on top for the rescue. Adding further to the complexity of this particular rescue operation, there was a large PVC-pipe-encased antenna bolted to the side of the pilothouse, reaching up about 40-

45-feet in the air, swinging wildly, "like a sailboat mast," recalled Branch, as waves from the heavy seas pitched that boat back and forth like an out-of-control rocking chair.

With Branch having to keep his aircraft well clear of that tall, whipping antenna, there was no way of safely placing his rescue swimmer on that boat. Throughout all of the on-site observations and rescue what-ifs, running rapidly through his mind, along with continuous input from his crew, the fuel state of his aircraft remained an ever-pressing reality and concern. And with remaining fuel now soon to become a factor, given their distance from shore, a rescue decision had, and all hoped the right one, with fuel and safety diminishing their options. After first trying to maneuver close enough to pick the captain up off the back of his boat, but prevented by that dangerously whipping antenna, following a quick consultation with his crew, Branch opted to execute a water rescue (i.e., the person(s) to be rescued must actually go into the water), rarely the first choice, especially with a boat that's rocking wildly, but dictated here by circumstances. Branch communicated the rescue plan by radio to the injured captain. He was OK with whatever method would get him to medical attention the quickest. The fishing boat's deckhand then helped his captain get into an on-board water survival suit. Once ready, the captain, no doubt hurting, somehow rolled himself into the water off the stern of his boat. That step completed, Branch maneuvered his hover as close to the rear of that boat as he safely could, then lowered his rescue swimmer down into the water near where the captain was floating, where the swimmer, then, assessed the injured man's condition, and prepared him for hoisting, as quickly and expertly as he could, under the illuminating beam from his helicopter above, in that pitch black night battling a churning sea.

When the swimmer signaled his crewmates that all was ready, the rescue basket was lowered from a height of about 30-feet, the preferred distance above the surface for hoisting. The swimmer then helped the boat captain into it, and the basket hoist was then successfully completed. Once safely inside the helicopter, Branch chose to leave the injured mariner in that basket to save time initiating the return flight. The swimmer was then quickly pulled up into the aircraft, by the standard method, with the hoist hook connected securely to his insertion harness. With all now onboard,

Branch and crew then headed directly for shore, with, thankfully, enough fuel to make it back to the field at Cape Fear, but very little more. Without the needed fuel to fly directly to a hospital, the overhead C-130 radioed shore to have an ambulance meet the flight at the field, and then transport on to medical attention.

The boat was fine, and the deck hand was able to get it back the 75-miles to shore. "We try to work with mariners when we have to pull them off their boats. We realize that if the boat is left without any crew, it's bound to get destroyed or stolen. Assuming someone(s) is left onboard, we'll typically have a small Coast Guard boat from shore go out and meet them somewhere. We'll also have them in radio communication with the Coast Guard Sector, so those on the boat can report every 30-minutes or so that they're still OK. And we'll coordinate with Sea-Tow if it ends up that they need assistance making it to shore," said Branch.

Branch did find out that the boat captain survived, but uncertain whether doctors were able to save the arm. Getting him off that boat by air was definitely the right call. "He was in bad shape," he remembered.

On the much lighter side, Captain Branch recalled an unexpected, and most unusual, side activity from back when he was in flight school. Renting a house there with his wife, one day a neighbor happened to mention that the local minor league hockey team was looking for a new person to play the role of mascot. Branch responded that, for a year at the Academy, he had been a team mascot, "so if they're looking for someone with college experience, have them call me," he said in an offhand joking manner, thinking that neighbor was kidding. The very next day, he got a call from the owner of that local hockey team indicating that he'd heard that Branch might be interested in the mascot position! Branch was, at first, convinced this was a prank call from one of his buddies in flight school. So, the caller said, let me try this again. He repeated his name, said again that he was the owner of the Pensacola Ice Pilots, and that he needed a yes or no answer. Realizing the call was legitimate, this time Branch replied, yes, he'd like the job. And so, with the permission of his Squadron Commander's approval (agreeing that this approval would be just between the two of them, so that when fellow commanders went to the hockey games, they wouldn't know it was Branch in costume), he was hired as the team's mascot, known as the "Iceman"! His costume consisted of a hockey outfit from the neck down, with a cartoon character headpiece, topped off with the old-school flight cap and goggles. Clearly not the ideal look to impress a date (or anyone else, for that matter). Branch's "Iceman" appeared in two to three games per week for about six-months. Remembering all the while that his demanding daytime pursuit was training to earn the wings of a Coast Guard pilot.

The Navy's Blue Angels are home-based in Pensacola and would often do out-on-the-ice promotions for the "Ice Pilots," who drew their name from this world-renown flying team. Before one game, Branch was on the ice sitting next to a Blue Angel pilot. After a brief greeting exchange, the pilot leaned over to Branch and said that he'd been told that the "Iceman" was actually a flight student, and asked if that was true. "Yes, sir, it is," said Branch, notably surprised by the sudden question. "Wow, that's awesome," replied the Angels pilot. Sensing a unique opportunity, Branch quickly thought to himself "if I don't ask, I will hate myself for the rest of my life." Not wanting to lose the moment, he leaned back toward the pilot and said: "Well, sir, if you ever need ballast in the back of Number 7, let

me know, I'm happy to do it!" Without hesitation, the pilot leaned back toward Branch and said: "What are you doing tomorrow?" Branch was stunned by what he'd just heard. He responded: "Sir, please don't mess with me, because I will drop dead right now if you're messing with me." The pilot let him know he was quite serious and to be at the hangar the next day at "zero-8."

*Coast Guard Flight School Student Lieutenant Marshall Branch flying in a Navy F-18 with the Blue Angels! (1998).*

"So, I was at the hangar at zero-7!" said Branch. He had phoned, in his words, "people I hadn't talked to since pre-school." Exaggeration, of course, but a clear indication of how very special he knew this opportunity would be for a student pilot, or anyone else for that matter. And his opportunity of a lifetime Blue Angels flight was "awesome." "They were doing an air show practice that day, and the lead solo swapped out his Number 5 jet for Number 7 which is the two-seater. Sitting in the back seat of Number 7, they did the full practice, so we were the opposing solos that do the crisscrosses, the high angle of attack, and all of that." The next few minutes were to even more significantly elevate this elite exhibition flight dream come true. About halfway through the practice, the Blue Angels pilot asked Branch how far along he was in flight school. Branch responded: "Just finished aerobatics, sir." To which his pilot said: "Outstanding…your controls!" At that totally unexpected instant, in the midst of this demanding and precise exhibition flight, Branch could only say to himself: "do what?" Just then, the opposing jet, next to them in the pattern, throws the nose down and takes off. "Well, you're never going to catch him unless you do something!" exclaimed the pilot to Branch. "So, I pushed the nose forward, and pushed the throttle up, and immediately I could feel my face distorting. I just never knew you could feel G-forces that way," he remembered.

After more time than he ever could've imagined piloting that elite jet through loops and barrel rolls, his Blue Angels flight host once again took the controls for the duration of the flight. But what an unexpected, fantastic experience for a Pensacola flight student, who of all things was doubling as a hockey team mascot, and just happened to be at the right place, right time, with the nerve to ask the question. "So, I got to fly a whole practice program with the Blue Angels, and the Navy yeoman even put in my logbook. So now it's permanently in my flight logbook that I have 1-point-3 F-18 time, and point-8 of that is First Pilot, which means I was at the controls! I am one of the only, if not the only, Coast Guard pilot with actual F-18 flight time," said Branch, describing that memory now, and after all these subsequent years of demanding and courageous Coast Guard flight, yet still, understandably, more than a little awed by that unique experience.

Jumping ahead several years, after giving up his brief brush with show-biz (the "Iceman") and, more importantly, earning his wings, along with the ever present water-craft rescues, no matter the Air Station assignment, and the narcotics interdiction role handled at some, Branch and his respective flight teams have been called upon to handle some highly important Presidential air-escort missions as well, generally conducted with little or no advance public awareness, and purposefully so. For Branch, this mission occurred primarily while he was assigned to Air Stations in Traverse City, Michigan and Los Angeles., working closely with the Secret Service, whenever the President was on the move.

"We do the security zone enforcement, and we also do the "huntsman" mission, which is where we are the airborne platform for the Secret Service agent watching over the motorcade. In that role, we do the site survey right before Air Force One lands. Once the air space is cleared out for Air Force One coming in, the Coast Guard helicopter will do a lap around the airfield, checking the perimeter with the Secret Service agent in the back, just double-checking to make sure everything's clear. As Air Force One's coming in for a landing, we are the only aircraft anywhere close by. So, I have been able to get 'on' Air Force One's wing as its coming in, and ride that in. Once on the ground, the President jumps right into his motorcade. When the motorcade pulls out, we lead it in the helicopter, flying about a mile ahead. The agent in our helicopter has the radio and he's calling down about any cars stalled on the side of the road or anything else that looks to be a hazard and wasn't pre-briefed. The Coast Guard becomes the visible airborne asset, although there are other assets in the air that aren't quite as visible," related Branch

And speaking of motorcades, while serving in Traverse City, Branch recalled flying with a 2004 Presidential one all across the Mid-West. As he remembered it, President Bush would stop at every town hall along the way. He would hold a big rally at each stop, then rejoin the motorcade, and head down the road again. Branch would land in a field next to where these town halls were being held and wait for President Bush to come out. "When word came that he was ready to go, we'd pull up in the air and lead the motorcade again," remembered Branch. This pattern repeated itself for several hours.

But on this particular day, the weather was starting to get bad. As it got worse, and the clouds grew lower and darker, they soon heard over their radio 'Aircraft Disengage,' that order given due to the increasingly poor flying weather. "At that point, then, there were only two of us left with the motorcade, our helicopter and Marine One, the helicopter that supports the President. As I recall, the motorcade was just pulling into Iowa. We were now following the Presidential vehicles flying just above the road. We'd slowed back to about 50-knots, so that we could see and avoid powerlines and things like that," said Branch. But despite the weather, his helicopter and the Marine aircraft stayed with the President's motorcade the whole time. However, with one unexpected operational change. Since the Marine helicopter didn't yet have the Coast Guard helicopter's advanced avionics, providing a welcome assist for bad weather flying. For safety, Marine One then began flying <u>behind</u> Branch's aircraft. Major credit to the Coast Guard!

Branch: "When we finally approached the airport (overnight stay location), we called up to the tower to let them know we were arriving. The tower responded: 'Well you'll just have to tell us when you land, because I can't even see the runway from the tower!' So, we climbed back up over the airport and all we could see were the end of the runway lights. We had to very progressively come in, and then we landed. The tower next said: 'You'll have to tell me when you're on the ground, because I still can't see you.'" Branch told the surprised tower controller that they <u>had</u> landed and were safely down. Then they shut down the aircraft and 'buttoned it up' for their Iowa stay that night. Soon a gentleman approached Branch and his crew, introducing himself as a Secret Service agent with the security detail for the President's motorcade. He then said to Branch: "Hey, nice job to you and your crew. You guys were the only air asset left. We kept looking up through the window of the limo wondering how in the hell are those guys still flying!!" Quite a compliment to Branch and his crew from a federal agent who, one can assume, in his line of high-stress work, was not normally or easily impressed.

A nice compliment, certainly, but given following a Coast Guard flight in far less than ideal conditions. Conditions that were, in fact, risky for Branch and his crew. Regarding the risk element that to varying degrees is almost always present, from the limited degree when shadowing a VIP motorcade on a clear day, to the extremes experienced with a nighttime rescue at sea in stormy weather. "We're (Coast Guard flight crews) Type-A personalities," responded Branch on the subject of flight risk. "When we get launched, we expect to complete the mission. But there are times when the level of the emergency doesn't warrant that level of risk. For instance, if flight conditions pose an extreme challenge, and the person(s) needing rescue is definitely <u>not</u> in a life-threatening situation, the risk-based decision may well be to either wait for the weather to clear or, if available, let a Coast Guard boat make the rescue," said Branch.

"We have lost aircrews," he continued. "We went through a period about twenty-years ago where we lost about six aircraft. It impacted us all, deeply. We're a small service, and aviation in the Coast Guard is even smaller, so we knew people who were on every one of those aircraft. It was a pretty dark time for Coast Guard aviation. I lost friends and mentors," remembered Branch.

As a result of those significant losses, policies were changed. "There's an old adage in the Coast Guard that says: 'You have to go out, but you don't have to come back.' It was after that time-period of dramatic loss when we officially got away from that. You <u>have</u> to come back, and as the aircraft commander, your job is to bring your crew back. So sometimes you have to make the call, where the safety of your crew is balanced against what it is you're trying to accomplish. And I think as a service, we've come a long way in formalizing that risk-management process," said Branch.

As an example of that change in risk-management policy and practice instituted after those significant losses, Captain Branch cited an example. "One of the rescue flights I lost a friend on was a case off of Humboldt Bay, California. A sailing vessel was getting torn up pretty badly. Terrible storm out there. Those on board were scared and injured. But in the end, we lost a helicopter in that storm. That crew did not come home. The sad irony is that boat made it back to shore under its own power!" From that experience, said Branch, "we learned a lot about us pushing the issue, because we want the rescue. Now with risk-management, evaluating the situation, not only before you leave, but as you get on-scene, and more facts become available, we continue to assess our risk."

"One of the things I'm the most proud of with Coast Guard aviation," continued Branch, "is that we have a concept called CRM: Crew Resource Management. The whole purpose of that concept is, if we hit the water, we're all going to go down together. So, every crew member has a stake in this game. Everybody's voice matters. And we train every member of our crew to speak up. Talk about risk, talk about concerns, talk about changes to the game-plan. Before the flight, and absolutely during it. We'll actually evaluate people on their ability to speak up. That's now become a part of who we are.

Co-pilots, we expect them to speak up and take the controls if the aircraft commander is incapacitated or doing something he or she shouldn't be doing. We expect the flight mechanics to speak up, and say for example, hey, you're below the altitude you briefed. We expect the rescue swimmers to speak up and say, for example, hey, did you guys see this aircraft behind us to our right? Bottom line, we expect everyone to speak up, because we all have a stake in the safety of our flight."

And, if there's any person in that aircraft, said Branch, no matter how new they are, and no matter how senior the aircraft commander is, if they don't speak up, then they consider this important, potentially lifesaving, crew safety measure to have failed. "So, we really harp on CRM (Crew Resource Management) in a big way in the Coast Guard, because we've lost crews, when someone should have spoken up, and didn't. Thankfully, today, I feel that we're in a much better place. Our risk-management processes, and our CRM, have kept a lot of crews safe. Because we do fly into bad stuff. Unfortunately, we don't get most of our calls on beautiful days!" said Branch.

On June 22, 2018, Captain Branch's two-year assignment as the Commander of Coast Guard Air Station Savannah, ending all too quickly, drew to a close. His one remaining duty was the traditional Change of Command ceremony, which was held in the Air Station's hanger. At that time, Captain Branch relinquished command to Commander Brian Erickson, who transitioned to Savannah from a tour at Coast Guard Headquarters in Washington, D.C. Presiding at the ceremony was Rear Admiral Peter J. Brown, Commander of the Seventh Coast Guard District, headquartered in Miami,

Florida. Among the many attendees, fellow Coast Guardsmen and friends, old and new, was Army Major-General Lee Quintas, Commanding General of the Third Infantry Division, Fort Stewart and Hunter Army Airfield, who recently returned from a nine-month deployment to Afghanistan. Also, in the audience that day was Army Lieutenant-Colonel Ken Dwyer, Garrison Commander at Hunter. Coast Guard Air Station Savannah is a tenant unit on the Hunter installation, which is proudly home to personnel from all five military service branches.

Captain Branch leaves the Air Station for a two-year assignment as the Coast Guard's Liaison Officer to the Navy's Fleet Forces Command in Norfolk, Virginia. "I'll be the lone Coastie, on a very big Navy base, working for the Four-Star coordinating operations, exercises, and mutual support," said Branch. "The Coast Guard actually uses the Navy quite a bit, both in law enforcement and, at sea, for the war on drugs. Their sensor capabilities are excellent. We will put specialized boarding teams onto destroyers and frigates, and once they detect something that needs to be intercepted, they will hoist a Coast Guard flag over the ship, shift authority to us, and their ship becomes, in effect, a Coast Guard Cutter, now with law enforcement capability. For search and rescue, major surge operations, like hurricane response, we'll use the flight-deck equipped Navy ships as 'lily pads' for fuel and logistics support (for CG helicopters). We'll also bring in the Navy's amphibian ships for command and control for major operations like Hurricane Irma in the Caribbean," he said.

Branch was asked what his most satisfying assignment in the Coast Guard had been. "This one (Air Station Savannah), no question" he responded without hesitation. "As a pilot, you want to go fly the missions. You want to do the rescues. You want to do all that stuff. It's just in your blood. But watching these crews here go through two historic hurricane seasons. Surging to Texas, to Louisiana, Florida, Puerto Rico. Flying into the teeth of those storms and saving dozens, upon dozens, of lives, each. I've got a first-tour guy who has more lives saved than I had in my whole career. Because he was down at Hurricane Harvey and the whole time he was there, he was just pulling people off of houses," remembered Branch.

"And I'll tell you, there's a part of you that wants desperately to go. You want to grab that helicopter and go. But there's something so supremely satisfying about watching your crews go out there and make it happen. The pride you have in them getting the job done. And that's what I've waited this long in my career for. And it's happened here. It's more than I could ever have expected. It's been amazing, said Branch.

*Deployment to Kuwait to support Coast Guard Port Security Unit 309, conducting Harbor/Port Security Missions at Kuwaiti Naval Base/Camp Patriot.*

If it happens that there'll be no more flying with your next assignments in the Coast Guard, what will you miss the most? "The people," replied Branch, again with no hesitation. "The reason why I went with the HH-65B helicopter is because it's the one that deploys on the backs of ships. I wanted to deploy with the crews. When you deploy with a unit, you create a much stronger bond than when you're in garrison. And I wanted to have those experiences where I deployed with the crew and really got to know them. We worked together to accomplish missions, to achieve tangible results, and that's something that will forever be with me. The crews that I deployed with (Eastern Pacific, South America, Antarctica), are the crews that I keep in touch with the most.

*The Branch Family arrived for a visit with Dad at work, just as a SAR case call came in! So, the kids got to watch while Dad and his crew "spool-up" the helicopter prior to launching on that mission.*

And I guess that's why this command tour has meant so much to me. I feel like it's been a two-year deployment with a hundred people! I've gotten to know each and every one of them and their families. I've been able to celebrate the good times, and I've been able to help them through the tough times. And I think that's what's made it really special," said Branch.

Looking ahead, there's something else that's really special, really important to him. "I'm still loving my job with the Coast Guard. I smile on the way to work. And as long as that keeps happening, I'm going to keep doing this until they tell me to stop!" concluded Captain Marshall Branch, with the same genuine enthusiasm, commitment, and caring that he's exhibited throughout his Coast Guard career.

## PERSPECTIVES ON LEADERSHIP

*What do you feel are the most important elements of a leader's character?*

"For me, I've always subscribed to the three "C's": Compassion, Candor, and Commitment. You cannot lead if you do not have compassion in your heart for those you are responsible for. Compassion allows you to put your people first, care for their needs and growth, and temper corrective actions with a sincere desire to learn what factors were behind any missteps. Candor, or honesty, is absolutely critical. Your people are smart. They will see through a façade. Be candid about your own shortcomings and mistakes, as well as those in your charge. It is important to temper candor with compassion, the combination of the two allows your team to know exactly how they are performing, while remaining confident that you have their backs and are supporting them to succeed."

"Lastly, you cannot expect anyone to be committed to a cause when they don't see that same level of commitment in you. Commitment is more than words. It is the actions you take to show the priority you personally place on the missions your unit is responsible for. If you say "safety" is a priority, you better live that mantra. If you say proficiency is important, then you better be getting as many reps and sets as you can. Again, military personnel are smart, and they pick up on hypocrisy, real or perceived, and that can undermine success. Commitment is the root of the phrase 'Lead from the Front.'"

*Please provide a memorable example that you've observed of exceptional character in leadership.*

"In the context of leadership, my memorable observation comes from Hurricane Katrina. Then Commander Bruce Jones (who retired as a CG CAPT) was the CO of Air Station New Orleans during the Hurricane Katrina response. As you can imagine, nearly every member of the crew of Air Station New Orleans suffered extensive damage to their homes and possessions. The Air Station was torn to pieces, all power lost.

"Despite it all, CDR Jones kept his crew motivated on the task at hand. His Air Station was the first to launch aircraft to rescue those trapped by the flood waters. And they continued to do so for the next 2 weeks straight. Around the clock operations, sleeping on hangar floors and in un-air-conditioned offices. CDR Jones checked on every member of his crew, and every transient crew that descended onto Air Station New Orleans (over 30 aircraft clogged the ramp of his 5-helicopter unit running SAR operations 24/7). I arrived 4 days into the operation as the senior member of the Critical Incident Stress Management Team, a team of aviation peers trained to help prevent PTSD in operators. I watched as CDR Jones inspired his exhausted crew, pulled every string he could to get FEMA trailers brought to the Air Station, so the crews could get adequate rest, and even managed to get a portable shower trailer brought on base.

"It was into the second week of the response that I was trying to find CDR Jones, to brief him on some crew issues, and I couldn't find the CO's FEMA Trailer. None of the crew knew which one was his, so I went to his office to try to find him (which was in the command building that had flooded and had an inside temperature well over 100 degrees). As I knocked on the door, I heard some rummaging and a quick "Come in". As I opened the door, I saw CDR Jones quickly pushing a bedroll under his desk. I said, "Sir, do you not have a trailer to sleep in?" He looked at me and shared that he would not sleep in one of the air-conditioned FEMA trailers, until every member of his crew was able to, as well (the trailers were trickling in about 4-5 per day at the time). That's when it hit me. "Sir, your crew doesn't know you are sleeping in here, do they?" He didn't answer, but he did ask what it was that I had come to tell him about his team.

"CDR Jones put his crew first, because he genuinely cared about them. But he didn't do it to fanfare. And he didn't even want them to know about it. Because he didn't want them to have the distraction of being concerned about their CO sleeping in a hot, moldy office.

"CDR Jones led the most tired crew I've ever seen, through the most historic rescue operation in our 200-year service history, saving over 12,000 lives and without suffering a single mishap. More importantly, every member of that Air Station New Orleans crew would have stormed the gates of Hell for CDR Jones. That's why CAPT Bruce Jones was the first person I called, when I was told that I would be commanding Air Station Savannah. Because he's the kind of leader I strive to be like."

*What do you feel are the most important traits or qualities of exceptional leadership?*

"I feel like this dovetails with the question about character. The two are inseparable. You can't be an effective leader without character, and the qualities that define character also define effective leadership.

"Aside from the character qualities, you could add traits such as communication and vision. A leader must have a vision for where the unit needs to go in terms of mission, professional growth, and performance. In concert with that vision, the leader must be able to communicate what is desired in a way that the crew can digest and adopt. Additionally, communication requires the ability to listen. A good leader knows how to listen to trusted advisers, to contrarians, and to the masses. Listening to trusted advisers enables you to gain critical insights for improving the effectiveness of your plan. Listening to contrarians forces you to think about perspectives other than your own. And, finally, listening to the masses allows insight into how your message is being perceived at various levels."

*Please provide a memorable example you've observed of exceptional leadership.*

"CDR Jones' example above is, without question, the finest example of leadership I have personally observed. That said, another example that struck me early in my career was that of a previous Commandant, ADM James Loy.

"Early in his tenure as Commandant of the Coast Guard, a particularly nasty public affairs crisis had developed. Just off the beach in Miami, a Coast Guard small boat had intercepted a tiny vessel with Cuban migrants attempting to make it to shore. At that time, the "Wet-Foot, Dry-Foot" policy required Coast Guard teams to try everything possible to prevent migrants from making landfall. In this situation, several migrants had jumped into the water in an attempt to swim to shore. As a news helicopter flew overhead, and as news teams on the beach recorded, the Coast Guard crews escalated through our Use of Force rules (our ROE) and ended up spraying the migrants in the water with pepper spray, to stop them from swimming further, so they could pull them onto the Coast Guard boat. The public outcry as this played live across the country was deafening.

"As a young Junior Officer, I fully expected the Commandant to either avoid the cameras by sending out a spokesperson, or to simply blame the boat crew for being too overzealous in their interdiction. Instead, ADM Loy went in front of the cameras and stated that they were following current Coast Guard policy, "My policy", and that he realized, due to this incident, that the policy is flawed. Repeatedly, he told the interviewers that he 'owned' the policy that caused this event, that the boat crews had effectively followed his policy, and that, going forward, he would change it. I was stunned. And impressed. The ADM protected that crew the right way, put the fire of public outcry squarely on himself, and fixed the problem!"

*How do you define courage? Is there a memorable example of exceptional courage?*

"I define courage as being scared, but doing what has to be done. I define exceptional courage as being scared s—less but doing what has to be done in a way that reassures those around you that the response you've chosen is going to work out. I've been blessed to see that done regularly in my job. Coast Guard aviators consistently go out in the worst weather, over the nastiest seas, to save people in exceptionally bad situations. Pilots, aircrew, and rescue swimmers exhibit bravery and courage so routinely, that it actually almost becomes routine.

"If pressed for an answer, I would point to an Air Station Savannah Rescue Swimmer named Garrett Downey. The aircrew was launched to respond to an emergency beacon 40-miles off Charleston. They happened to have a Go-Pro and filmed the entire rescue (https://www.youtube.com/watch?v=KxIiN559Jms). At around minute-6, you can see Garrett swimming fast up to the overturned boat, pulling himself aboard, and talking calmly with the survivors of the now exceptionally distressed vessel.

"What you don't know, is that as soon as he entered the water, a large Mako shark started chasing him and made several aggressive runs at him, trying to bite him. Garrett swam fast, punched back, and continued to the vessel to rescue those boaters. Once on the overturned vessel, you would never know he was being chased by a shark. Later, the survivors would enter the water to be picked up by another boat that had pulled alongside. Garrett stayed in between the shark and the survivors to make sure they all got on the rescue boat safely. That's being scared s—less, but regardless, doing what has to be done, while keeping those around you calm while doing it!"

*What have you most valued about your military career? What has your military experience meant, and, importantly, given back to you?*

"First and foremost, I value the continuing career service to this country that I love. I consider myself incredibly blessed to have been in a position to help those who needed to be rescue, and to have contributed to the defense of our nation during my time in uniform. It has been an honor and a privilege to be able to perform the myriad types of missions the Coast Guard achieves every day, from rescuing those in peril, to interdicting illegal drugs, to protecting our environment, to enforcing homeland security, and supporting homeland defense. Every single day of my 20+ year Coast Guard career-to-date, I was able to directly serve my country and my fellow citizens. That has brought me incredible pride and purpose.

"A very close second is the brotherhood/sisterhood of military service. I've been blessed to serve with some of the finest men and women that I have ever encountered. The friendships you develop during a military career are every bit as close as family. Military service has given me mentors, friends, and experiences, that have both made me a better person, and enriched my life. I've grown as a person through the challenges of leadership and responsibility, with sincere thanks to the tireless efforts of some very talented Chief Petty Officers, senior officers, and junior enlisted. I've seen the best our nation has to offer.  And I continue to be inspired by the endless number of truly great Americans, talented, dedicated people, of all ages, races, and backgrounds."

**UPDATE:** From 2018 – 2021, Captain Branch was assigned as the Coast Guard Liaison Officer to the Navy's U.S. Fleet Forces Command HQ in Norfolk, VA. U.S. Coast Guard Captain Marshall Branch retired from active duty on September 1, 2021, after 26 years of commissioned service.

Since 2021, Marshall Branch has been working for TQI Solutions, Inc., a service-disabled, veteran-owned small business, supporting IT modernization for the U.S. Navy. He is currently the company's Deputy VP of Operations. He and his wife are living in Leesburg, VA.

# CHAPTER 4

## "I Could Still Feel My Heart Beating, So, I Knew I Wasn't Dead" Green Beret & Silver Star Recipient Colonel Ken Dwyer United States Army

Colonel Kenneth M. Dwyer was destined for a military career. Born in Tampa, Florida, at MacDill Air Force Base, but "'raised everywhere." His Dad was career Air Force, with successive moves around America, largely in the Southeast, while young Ken went through the usual phases of growing-up. Between his dad's lengthy service, and his Granddad's career in the Navy, there was little question that he would eventually enter the "family business," military service, of course, specifically the United States Army, "to round out the family (branches)," he said.

When it became time for college, with an Army ROTC Scholarship in hand, Ken chose Furman University in Greenville, South Carolina, for four basic reasons: (1) Furman had an Army ROTC program; (2) his sister was already enrolled there; (3) it was close enough to his parents, now back in Florida, to drive home during breaks; and, significantly, (4) it was very close to his Grandparent's home, where he could escape for a home-cooked meal and do laundry on the week-ends!

While still an ROTC cadet, he was afforded the selective opportunity to attend and complete both the Airborne and the Air Assault courses, during junior-year summer training at Fort Benning, putting him two important qualifications ahead of many of his fellow cadets. Dwyer graduated from Furman in 1998, now commissioned as an Army Infantry Second Lieutenant. Looking back at that first phase in his career, he recalled that he "loved the infantry, absolutely! And of course, infantry guys will tell you there are only two jobs in the Army: the infantry, and then everyone else who supports the infantry!"

His first posting was back to Fort Benning for the Infantry Officer Basic Course, followed directly by Ranger School. "It's expected of Infantry Officers to have the Ranger tab before they get to their first unit," recalled Dwyer. "It doesn't always happen, but it's expected. If you show up at an Infantry unit, as a Lieutenant, without a Ranger tab, it's frowned upon."

Ranger training consists of three sites: Fort Benning (initial training), then up to Dahlonega, Georgia (mountain phase), and finally down to Fort Walton Beach, Florida (swamp phase). When asked what he remembered the most about Ranger School, without hesitation, Dwyer said: "Being cold, hungry, and miserable! That's the surface level stuff you take away from it and remember. But what you really learn in Ranger School is how to continue to function when you're cold, wet, tired, and hungry. More importantly, how to lead others who are cold, wet, tired, and hungry. How to get the most out of your people, even when they're at their lowest."

Ranger school successfully completed, his next step was a wise one, and one that would prove to be of life-long importance. Second Lieutenant Ken Dwyer married Jennie, his finance, and Furman classmate, in January of 2000, in Greenville, South Carolina. Said he, while speaking to Georgia Military College cadets many years later: "The most important decision you will ever make in your life is who you're going to spend the rest of it with. I've made two good decisions in my life and a ton of bad ones! I picked the perfect woman to marry, which was good decision #1. And #2, I joined the United States Army. Those two decisions set me up for where I am today." (via unionrecorder.com/Gil Pound). With Ranger tab proudly affixed, he was then off to his first unit stationing, joining the 101st Airborne Division ("Screaming Eagles") at Fort Campbell, Kentucky, for a three-year tour there with this legendary fighting unit. Shortly after his arrival, he was promoted to First Lieutenant.

During his time with the 101$^{st}$, he did his first seven-month tour in Afghanistan, back at the very beginning of the war. "Most of our unit got to Afghanistan in December of 2001. We were the first conventional forces to enter the war. Obviously, there were Special Forces guys there on the ground prior to the 101$^{st}$'s arrival. Which, incidentally, became my big reason for transferring over from Infantry to Special Forces. The impact they had on the battlefield prior to us even getting there was considerable," remembered Dwyer.

Following that deployment, once back at Fort Campbell, he sought out the Special Forces recruiter on post, advising him of his desire to transfer from Infantry to become an "SF guy." He was told that he had missed the window to submit his application paperwork while he was deployed to Afghanistan! Meantime, Dwyer, now a newly promoted Captain, received orders to return to Fort Benning to attend the Infantry Captain's Career Course. While there, he learned that the Special Forces application window had been re-opened! Without delay, he submitted his paperwork, to both transfer from the Infantry, and for acceptance into the Special Forces Assessment Course.

So, on temporary duty status, it was back to Fort Bragg for a 24-day rigorous assessment of his capability to succeed in the next phase, the Qualification Course, which is the final requirement to obtain the Special Forces tab. As might be expected, given his successful career performance to-date, Captain Dwyer performed very well in the assessment phase, and was "picked up" to attend the SF Qualification Course (Q-Course), also conducted at Fort Bragg, a demanding year+ in duration (longer for medics and certain other specialties), to, hopefully, finally fulfill his ultimate developmental Army goal of becoming an "SF guy." But first, he had to return to Fort Benning to await his orders, prior to heading back up to Fort Bragg to begin the Q-Course!

After successfully completing all required elements of the SF Q-Course, the first time through (only about 50% of candidates achieve that distinction), Dwyer was assigned to the 3$^{rd}$ Special Forces Group, allowing him to remain at Fort Bragg. Told by his battalion commander that he'd be put on the dive team, which at the time, was the only slot available, he was then sent to Key West, Florida, to attend the six-week Combat Dive Qualification Course, where, among other tasks he would find himself physically tied-up, well below the surface. "It was a great time, but maintaining focus under water when you can't breathe is a challenge !!," recalled Dwyer.

*Capt. Ken Dwyer, Team Leader, ODA-325, 3rd Special Forces Group, Afghanistan 2005.*

Upon completion, it was back to Fort Bragg where he became a 3rd Specials Forces Group Detachment Commander of Operational Detachment Alpha 325. Beginning in 2001-2002, SF life consisted of continuous rotations to Afghanistan. "You would spend six-months at home, then six-months deployed, then repeat the sequence. That was pretty much the life of team members back in those years," said Dwyer.

With two combat deployments, by then, behind him, Dwyer's third Afghan tour began in August 2006. After arriving at their designated Fire Base ("Cobra"), his team's first mission 'outside the wire' was to identify key terrain features that Dwyer assessed to be 'strong points' and then to establish control measures to protect the base and surrounding area. "The idea was to basically build 'white space.' What you wanted to do in the area, ideally, was to ensure that we had freedom of maneuver, as well as the Afghan people and the Afghan government, and the insurgents did not. Building 'white space' requires moving out from where you're centrally located to establish those 'strong points' and security positions, to keep the enemy from coming into those defined areas, whenever our troops return to the Fire Base," said Dwyer.

But how do you then keep that 'white space' secure? Dwyer responded: "You identify key terrain, but we don't have the assets to defend it, so you in-place your Afghan counterparts and forces. You have Afghan assets. And it's not just Afghan National Army assets. You also have Afghan security forces inside different villages. So a lot of the job at the time was moving into a village, and working with the Afghan governor, Afghan locals, and Afghan security forces, whether they were police or border guards or whatever, and you work with them to develop a security plan. Because it's not just us over there fighting our own war. It's us fighting against the insurgents with the cooperation of the Afghan government."

On this, his third Afghan deployment, and only his third day in-country, at Fire Base Cobra, on just his second movement out and away from the comparative protection of that fire base, this time Dwyer took his combined team in a different direction from the day before. With him this time, was a force of about 12 U.S. Army soldiers (mostly SF), and around 30 Regular Afghan Army soldiers. Insertion into this new area would be by means of six Army diesel-powered Humvees (not yet the later up-armored version), transporting the U.S. soldiers, while the Afghan troops would follow, riding exposed in Toyota pick-ups, as was their practice back then. The group was, again, heading out on a combined combat reconnaissance patrol, to once again try to identify key 'strong points,' so that additional protective control measures for the surrounding area could be established. "Since we had just recently arrived at that Fire Base and we needed to get 'eyes-on' the various terrain parcels that we 'owned,'" remembered Dwyer. "This time, we knew there were some enemy fighters said to be in the area. The intel that we had, estimated perhaps 20-25 Taliban insurgents that we might expect to encounter. Should that happen, we were confident that we had sufficient assets on hand to take care of that enemy number with no issues."

However, when Dwyer and his soldiers rolled down into a remote valley, amidst "pretty brutal, mountainous high desert" terrain, they quickly discovered that the insurgent estimate was faulty. Really, really faulty! Rather than a manageable number, in the 25 fighters' range, the enemy troop-size turned out to be closer to 125! Worse, that over-sized force was dug-in well above the 'bizarre' (small valley center with clustered mud-hut shops along a dirt street), and "we kinda got stuck in a bad ambush late that morning," he recalled, understated, in a Special Forces leader kind of way!

Regarding enemy presence in an area, "you can pick up some indicators most of the time. Not all of the time, but most of the time, that you're gonna have contact soon," said Dwyer. "And we did pick some up, but we thought it was a force (based on the advance intel) that we were absolutely gonna to be able to easily overcome. But once first contact was made, it became pretty obvious that it was going to be a much larger problem than we had anticipated."

On the outskirts of that small village, their lead Humvee was immediately engaged with RPG's (rocket propelled grenades), coming at them at a high rate, along with small arms fire, to include several machine guns firing from the enemy's dug-in positions above the 'bizarre.'

From the initial burst of that firefight, the lead vehicle, a U.S. Humvee, was immobilized. Captain Dwyer was in the second Humvee in the Allied line-up. So Dwyer's team began maneuvering to get to that lead vehicle, which carried several Americans, their condition unknown at that time. Sadly, several of the Afghan soldiers, riding farther back in the bed of those open trucks, were killed right away in the initial insurgent volley.

"We knew we had to get in there, recover that lead Humvee and get it out of the kill zone. You make an attempt to have a higher volume of fire than the enemy, so you can gain the initiative and then recover that vehicle, "said Dwyer. "But unfortunately, with the level of fire the enemy was pushing out, there was really no way we could gain the advantage."

At that point, Dwyer and his team maneuvered their vehicles through different sites, firing as they went, seeking positions where they could take at least some cover and engage the enemy. The vehicles they were in at the time, mostly Humvees, offered some protection against small-arms fire, but not the enemy's rocket-propelled-grenades (RPG's), in those yet to be up-armored Humvees.

Now confronted with far greater enemy numbers, what kind of assets did they have to effectively defend themselves and return fire? "We had turret guns, 50-calibre machine guns, Mark-19's, and Mark-47's (with 40-millimeter grenades/chain-fed/mounted atop some of their vehicles). And all of our vehicles normally had a rear-mounted 240-machine gun, which is what I was manning at the time. So, we did have firepower, to include everyone's personal weapons," said Dwyer.

He and his Air Force combat air controller had just gotten into the back of their vehicle (GMV), where the air controller would attempt to make radio contact with any assets flying near-by, in hopes of calling-in immediate air-strike fire power from above against the enemy fighters. Meanwhile, Dwyer was returning fire with that rear-mounted 240-machine gun.

The date was August 19, 2006. With both men exposed in the rear of their vehicle, with little to duck behind, and with enemy RPG's coming in all around them, it wasn't long before the air burst from one of them exploded close to the turret of Dwyer's Humvee. There, in that instant, with shrapnel spraying widely, hot pieces of metal slammed into his body, from head to waist, wounding him severely. Sadly, also hit by the blast, positioned as he was, near Captain Dwyer, his combat air controller took direct shrapnel hits to his head, killing him instantly. As it turned out, despite the intensity of the battle, while other of our troops were wounded, the team's air controller was the only American who perished that day.

At that moment, the problem and priority then quickly shifted from the lead vehicle being disabled, to the RPG that had impacted his vehicle and Captain Dwyer himself. Just prior to the hit, "I was actually helping the Mark-19 gunner on our vehicle reload, so I had a can of Mark-19 rounds in my left hand, which I was holding up in the air to give to the gunner, who was going to reach down and grab it and load it into the turret gun, when that enemy round (RPG) came in," remembered Dwyer vividly. He had been standing behind the turret, in the flat-bed portion of the vehicle, exposed to enemy fire, as he tried valiantly to help the gunner keep up the defensive firing. The turret gunner was also hit in the face with some shrapnel, as well as suffering a concussion from the force of the close-in blast, although his body was largely protected by armor plating surrounding his turret gun. To add to the gunner's injuries, as he quickly dismounted from his vehicle turret, he took an Afghan bullet in his arm.

"They always tell you, and you never believe it until it happens to you, if you can see a round coming in, you're OK. It's the one you don't see that's gonna get you," recalled Dwyer. Preoccupied as he was with trying to help his gunner reload, reaching those vital new rounds up to him, Dwyer's left arm was fully extended in the air, when "the one he didn't see," got him, and in an instant, changed his life forever.

That shrapnel blast "hit me and took my left hand off, along with metal shards in the face, neck, and the right arm. The flash from the RPG initially blinded me, and the explosion from it deafened me." He still well remembers, in the midst of a gunfight, at that fateful moment of impact, "everything went black and silent."

And what he recalls that was so strange about that shattering instant, as severely wounded as he was, with the very real possibility of dying from blood lose, right where he fell, there on the rear of that vehicle, as the intense battle raged on around him, unexpectedly, Dwyer never did completely lose consciousness.

He does remember the first thing that went through his mind at that moment. "Holy crap, I just got shot in the face and I'm dead. My next thought was anger. I was really pissed off. I just got shot in the face by an Afghan! Then I thought, I'm a Bible-believing man, so at least I know I'll be taken care of, I know I'm gonna be OK, fully expecting, at that moment, to float off to Heaven, or whatever happens next," recalls Dwyer, as his thoughts in those very first seconds continued rapid-fire.

But "floating off to Heaven" didn't happen. And Dwyer thought "that's not good. But it was at that same moment, that I started to feel, not hear, but feel my heart still beating in my chest. Oh, so I'm not dead. I could still feel my heart beating. Well then, let me get back up and keep fighting."

At this point, while realizing that he was, indeed, still alive, an assessment that he may have thought took some time, but in reality, took only seconds, his mind now began permitting him to realize the extent of his injuries. "I could tell that my left eye was gone. But in those moments, I was able to open up my right eye, and could tell I had a ton of blood and stuff on my face. I couldn't move my right arm because of the damage from the shrapnel. But I was able to kind of peel my left arm across my body and see that I didn't have a left hand anymore, that it had been blown off, and I just had two bones sticking out of my wrist," remembered Dwyer, still all too vividly.

And it was that sight, that visual of his hand completely gone, and only two wrist bones left sticking out, that would be the memory toughest for him to forget. "This is the terrible part of the story to me, personally, because it's seared into my brain. That moment, where I opened up my right eye, and brought my left elbow up in front of my face, and I see those two bones and all the charred skin around the edge of those bones, that visual remained stuck in my brain. That image happens, still to this day, but thankfully not as often as it did three or four years after the incident. When I lay down in bed at night, my left arm will pop up in front of my face, and I'll just stare at my limb, and I can visualize the bones sticking out of my wrist. And my wife has to reach over and push my arm down. And then it pops back up, and she pushes it down. Over and over. You're just kind of stuck in that moment. It doesn't happen now nearly as much as it used to, but it still pops in there every now and then. It's just one of those things in your life that is so surreal. And so life changing. Such an emotionally significant event, that it just sticks. You get past it, but it's always there," said Dwyer.

Turning back to the immediacy of the RPG blast and his grievous injuries, Captain Dwyer remembers how fortunate he was to have some incredible fellow Green Berets on the scene, in the midst of that firefight, to help him, and, frankly, to save his life. But now, we're at the moment, likely but a matter of seconds, when the pain finally set in.

"When it (the trauma) first happens," he remembered, "it's just such a shock to the system. There wasn't any pain right at the beginning. But once I realized that I'd been really messed up, that's when I felt the pain begin. You hear people talk about your body can only tolerate so much pain, and then you just pass out from it. That's total BS, in my opinion. Honestly, I was begging to pass out from the pain. It was hideous. It was terrible. It was worse than anything that I could have ever imagined," he said.

Now, with his very life on the line, what about the immediate danger of bleeding out from the severed left wrist? Although gone from the blast, as Dwyer explained it, it seemed, at least initially, to be "seared off." He remembers being told much later, that although it may bleed some, with the body's protective mechanisms for a significant dismemberment, the immediate reaction of the blood vessels is to contract and pinch. But once the vessels begin to relax, the larger, far more dangerous blood flow begins.

Surprisingly, at least immediately, it was not blood loss from his sheared-off left wrist that was the most threatening to his survival. Unexpectedly, it was his right arm. "A piece of shrapnel that went into my right armpit that just nicked my brachial artery. When the (Special Forces Team) medic got to me, he told me that my body was kind of pinched over my arm, and when he laid me out to get a good assessment, peeling my body off my right arm, arterial blood squirted out of my armpit, three-feet at a time, quickly covering my medic from head to toe," said Dwyer. So, it was the right arm that was actually the major bleeder, and one they had to worry about and tend to first.

The medic, who quickly put a tourniquet on Dwyer's right arm, had actually been the second person to get to him after the blast. The first soldier found Dwyer slumped down around the turret. So he grabbed Dwyer and rolled him over, at about the time Dwyer had opened his right eye and was processing what all had happened to him. "I remember looking at his mouth, as he yelled to other responders: 'Dead Guy'!! I didn't hear it. I just saw it in his mouth: 'Dead Guy.' Then he just rolled me back down on the bed of that vehicle and moved on to the next casualty! That's another one that stuck in my brain. So, I remember waiving the bones of my left hand around in the air, unable to talk because of the shrapnel in my face and neck, whispering as loud as I could, 'I'm still alive. I'm not dead yet."

Fortunately, his medic did find him, badly wounded, but very much alive. He considers that medic, by the way, to be "an absolute rock star, the best man I've ever met in the entire world." And, also fortunately, his SF team sergeant got to Dwyer at about the same time as the medic. "Both absolutely amazing people, and I wouldn't be here if it wasn't for them," he said, "as the team sergeant went to work on my left side, while my medic (also his work-out partner) went to work on my right." It's important to note, at this point, that the intense firefight continued all around them, as these two incredible soldiers did their best to apply emergency treatment, all the while ducking down as low as they could to avoid being hit themselves.

In the midst of a chaotic battlefield situation, Dwyer's medic didn't recognize him at first, but when he did, as he shared with Dwyer much later, "he froze-up, he was 'freaking out," realizing this was his work-out buddy, his close friend, and not just another American soldier in peril. But my team sergeant remained calm and was able to, with his voice, calm the medic down, so he could do his job. That senior non-commissioned officer held it together for everybody," recalled Dwyer. And thankfully so, at this ever so critical time in Captain Dwyer's attempt to survive.

"So, the team sergeant put a tourniquet on over here (left arm to stop the bleeding from hand loss), and to help hold my neck together (also a major shrapnel strike), while the medic was dealing with the arterial bleed from my right armpit," said Dwyer. The medic had taken a specially developed battlefield bandage, with compounds within designed to promote blood-clotting, and pushed it hard against the artery wound. He then put a tourniquet on top of that bandage, hoping that those two methods would combine to stop that pumping right-side blood loss before it could take Dwyer's life. While stopping the immediate blood loss problem, a mandatory step, the downside was that it also stopped the blood flow to his right hand, and for an extended period, with potentially catastrophic implications, to be revealed later on.

Once stabilized the best they could in the field, with that explosive firefight still going on all around them, staying as low and as shielded as that precarious situation allowed, carrying him just as quickly as they could, in the first step toward his hoped-for evacuation, Captain Dwyer was placed in the larger bed of one of the partially shielded Afghan Toyota pick-up trucks, carefully lying him down alongside wounded and dead Afghan soldiers.

At that point, as a stark indication that, despite his severe trauma, blood loss, and extreme pain, still well in the midst of a very dangerous and intense combat event, the Captain had somehow retained his sense of humor. So, when the wounded Afghan soldier lying next to him, in the bed of that truck, turned his head and made eye contact, Dwyer remembers whispering to him: "This place is dangerous." An observation that could well be the understatement of the entire Afghan conflict! And adding to the strangeness of that moment, he had actually whispered those words, <u>not</u> in English, but in Farsi, the Afghan language! Under such strained, and intensely painful circumstances, his bi-lingual clarity of mind was amazing. And in hindsight, as he now remembers and freely admits, given the noise and chaos all around them, the one statement he had somehow managed to whisper, in Farsi, with his neck damaged and his voice gone, was darned funny!

With Dwyer and others lying in the back of a pick-up, members of his combined team were able, at last, to pull back out of the immediate area of fire, managing to maneuver toward relative safety, just behind a slight ridge. An Army medivac helicopter had been summoned and was able to land far enough from the fight to, hopefully, avoid being hit as well. Dwyer and the others, both dead and alive, were then transferred to the medivac and flown to the nearest Field Surgical Unit (+/- 30-minute flight). Along the way, as Dwyer now knows, after reviewing memories of that day much later with his combat medic friend, that medic was offering him encouragement throughout that flight to still more emergency medical treatment. He told Dwyer: "Hey, you're gonna be alright. You're gonna be home to see the Dolphins get beat by the Steelers in a couple weeks." Words meant to lift the spirits of his badly wounded team leader and friend, in hopes that he could hear them.

Recall the tourniquet on Dwyer's right arm to stop the pulsating arterial blood from stealing his life. A right arm that had taken extreme shrapnel damage, on top of that to his artery. And, a right hand that, because of the tourniquet, had been deprived of blood flow for some time now. Dwyer was somehow aware enough of that to be worried about the then-strong possibility of losing that arm, along with everything else he had lost.

Arriving at the Surgical Unit, they carried him off the helicopter on a stretcher. As good fortune would have it, another medic appeared, assigned to that medical facility, and, with amazing luck, he was also one whom Dwyer knew. That awareness, that relationship, would, several minutes from then, turn out to have a miracle impact on the rest of his life. As Dwyer recalls: "He comes up to me, and grabs me, and says, hey, are you OK? I don't remember this, but he told me later, that I looked up at him, and repeated four times (recall Dwyer could only whisper because of all the shrapnel in his neck and face): "My blood type's A-positive, and don't let em' cut off my right arm!" And then they sedated him.

That hospital medic friend scrubbed for Dwyer's surgery, another stroke of good fortune. And that's because the female Army doctor who was operating on Dwyer, would end up concluding, recalled the medic: 'In order to save his life, we're going to have to cut his right arm off. It's had no blood flow to it for over an hour. It's dead anyway and it's mangled (from considerable shrapnel hits), completely mangled, so we're going to take it off.'

At that point, the surgeon started coming at the sedated Dwyer with the bone saw (confirmed later by both the surgeon and the medic!). But his medic friend quickly stood up and said: "No. ma'am, I'm not going to let you do it. I am not going to allow you to cut his right arm off." Let that scenario sink in. NCO medic effectively 'orders' the Officer Surgeon not to proceed with her best medical judgment. The surgeon could have easily reprimanded the NCO for being out of line. But she didn't. Instead, she considered that veteran medic's firm statement, and responded: "Well it's impossible to save it, but I'll try." A very critical turning point in Captain Dwyer's future. And for that, we have two 'heroes' to thank: That stand-up medic who openly disagreed with an officer on behalf of the wishes of his friend, and a compassionate surgeon, who, in that particular instance, was willing to put professional judgment and personal feelings aside and begin working to repair and hopefully save that arm.

So, she put down the bone saw and went back to his right arm. She sliced it open further, cleaned it out, and began sewing it up. She told Dwyer, much later, that she began sewing up muscle to muscle, even though she had no idea if it was the right muscle. She continued, then, by re-attaching artery to artery, and vein to vein. She admitted to him in that personal conversation, that she had no idea which was the correct pairing, because his arm was so badly mangled. It was such a mess that she had done the best she could to sew stuff back together.

Returning in real-time to that field operating room, she finished with his arm, bandaged it up, and was putting her instruments away. Then she said to the medic, knowing it was very likely to be impossible, but just for personal curiosity, she was going to check him for a pulse. She placed her thumb firmly on his right wrist. A pause. And then, miraculously, there it was, a faint pulse!

As Dwyer recalls, it did turn out to be many weeks before the doctors were certain that he would be able to keep his arm. What an amazing set of circumstances, and eventual victory. He had already lost his left hand. Just imagine how his life-struggle would have been further compounded by the loss of his entire right arm. While he didn't "float off to Heaven" as he had initially assumed, given his life-draining injuries, slumped helplessly on the back of that Humvee, there can be no question that, with all the right people, in all the right places, God Almighty was with him in those critical moments that day, and many others to follow, as he lay there in that remote combat operating room, now so far away from his team, and from home.

And Dwyer would find out much later, in talking with that talented Army surgeon responsible for not removing his right arm, that while he was still on that operating table, she had also needed to bone-suture his trachea back together. From the shrapnel strikes to his neck, Dwyer's trachea had been split into four pieces (little wonder his voice wasn't working). Although she was a bona fide ER doctor, she admitted that she had no idea how to put a trachea back together. So, she literally "scrubbed out of surgery," went to the nearest computer, pulled up a database to research how to do this delicate operation. Do recall that this was Afghanistan, in 2006, when doing searches on-line was far more involved than it would be today. But, amazingly, she did find the surgical instructions, went back into the operating room, and completed the delicate and critical surgery on Dwyer's throat. With his voice in such great shape today, one would never suspect it had taken, by then, a likely-fatigued, but still dedicated and clearly skilled trauma surgeon, with an on-line assist that, technology-wise, now seems like light-years ago, to restore his vocal abilities for a lifetime.

Following all of that emergency surgery, including stabilizing attention to his left wrist, where the hand had once been (cleaning out the severe wound to prevent infection and wrapping it), he was transferred to a special medically-equipped and staffed Air Force aircraft for a flight to the military's hospital at Landstuhl, Germany, usually the next stop for American service members wounded in Middle East fighting. Landstuhl Regional Medical Center is operated by the U.S. Army and the Department of Defense, and is the largest United States military hospital outside the continental U.S.

Dwyer was at Landstuhl for about two days. He has little memory of his stay there, as he was under continuous sedation. That is except for a few moments when his Battalion Commander, who had been delayed from arriving at Dwyer's Special Forces unit in Afghanistan, chose to delay a bit longer and detoured to Landstuhl to see him. Once in Dwyer's room, the Commander asked that Dwyer be brought out of sedation so he could hear him.

So, the medical staff began to bring him slowly back to semi-consciousness. Dwyer does remember his 'boss' telling him he'd done a good job, that he was going home, and that everything's going to be OK. He also recalls what happened next. "I remember trying to get up, and tell him to send me back in. Let me go back to be with my guys. They (medical staff) had to immediately begin putting me back under, because I started to try to sit up and get out of the bed, with all of these tubes sticking out of my neck!"

The next step for Captain Dwyer, via nearby Ramstein Air Base, was the long flight back to Walter Reed Army Medical Center in Bethesda, Maryland. "That first month I was there is so hazy, because of the amount of drugs I was on. Constant morphine drip. In and out of surgery. It felt like every day they were prepping me (for more surgery)." On that note, from start to finish, Afghanistan to Landstuhl to Walter Reid, he estimates that he endured at least 50 surgeries(!), addressing his multiple wounds.

Jennie Dwyer had received notification about her husband's severe injuries in combat, and when he would be transported to Walter Reed, so her dad met up with her at Fort Bragg, and together they drove to the medical center in Maryland. When she saw her husband for the first time, so badly injured and hooked up to about every medical device possible, it was, no doubt, an incredible shock, despite having been pre-briefed by his doctors. But for the Captain, her presence provided a much-needed emotional lift.

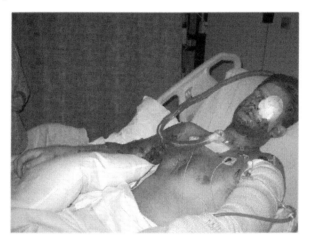

*Capt. Ken Dwyer, Walter Reed Army Medical Center, September 2006.*

Thinking back, again, to that early time at Walter Reed, "all I had that first month were just visions in my head, and hallucinations, honestly. Crazy things, like I'd go from being in a firefight one minute, and the next minute, I'm lying back in a hospital bed in Walter Reed. All of those noises and sounds that would go on in the hospital, to me, sounded like bullets flying past my head. I'd be lying there in bed, with these constant hallucinations of being back there in the firefight, but not being able to move, not being able to respond, and knowing that I was really messed up," he recalled.

Soon after his arrival, among the damage they focused on, of course, was his left eye. His doctors weren't sure they could save it. "Shrapnel went through my eye and deflated it, but it wasn't gone, it was still in there. So, they (doctors) would cover up my right eye, and then flashlight into my left eye, and tell me to nod my head (if I could see it). That's all I could do. I couldn't talk. I had a trach, a feeding tube, and everything. I couldn't see the light, but I wanted to believe I could see it, so I'd nod my head a couple of times. The doctors thought it was amazing, since even though the eye was completely deflated, I could still see out of it. But I was just faking it. Eventually, the Doc came in one day and said the same thing, nod your head if you see the light. So, I started nodding my head. Then he told me that he hadn't shined the light yet! That's when they went in and pulled the left eye out," Dwyer remembered.

With all the shrapnel damage to his neck and face, he couldn't eat, because of the holes that went from his trachea to his esophagus. Until they healed, he had to be on a feeding tube, an ordeal that lasted a full month. But in his memory, that wasn't the worst part. For Dwyer, a seasoned Special Forces warrior, the very worst was being so incapacitated that he couldn't even clean up after relieving himself, the most private of moments. "I had a bed pan and I'd s**t on myself, and I couldn't do anything. I'd just lay there. And you had to have someone come over. To be trapped in your own brain, while conscious and coherent enough to realize that you're that bad off. It was awful. It was hideous. And then to not know if I'm ever gonna get better. I think I am, but who knows," he said.

Early on, little triumphs really mattered, as he worked as hard as he could each day to recover. "I looked over at my left hand, it was gone, but I knew that. I couldn't see my face, but I knew something was messed up with it. In my head, I pictured that I had no face left. Then I looked up at my right arm and it was suspended in the air, and I could see my hand.

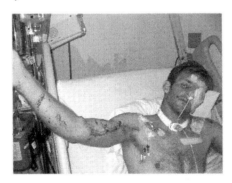

*Capt. Ken Dwyer, Walter Reed Army Medical Center, two weeks later in September 2006.*

So, I said to myself, OK, move your hand. I tried as hard as I could, and I couldn't move it. I thought, oh, that's awful, I can't even move a finger. So, I said to myself, now really focus and just move one finger. I tried really hard, and then I got a little twitch out of my finger. And I was just thrilled!! Like, over the moon. Because I was, like, yeah, I got a little finger wiggle," remembered Dwyer. Realize what the start of a vital breakthrough this was for him, since he would, hopefully, be depending fully on the use of his right hand for the rest of his life.

And the breakthrough continued. "Trapped in my own brain and lying there in my bed, I remember my goal in life was to try to wiggle my finger even more. And I did. Eventually I got to where I could wiggle it much better. And then all of a sudden, I thought, well, why not try to touch my thumb. And so, I would just lay there in the bed and touch my thumb, over and over. And then, eventually, I could wiggle two fingers. And I'd go back and forth (between the two). Understand, this took weeks! I'd just lay in the bed and touch my fingers. And, eventually, I could touch three fingers. It took me probably three months before I could touch my pinky (finger). I've got a ton of nerve damage, and three of my fingers are still, today, like pins and needles. I can't really feel them. They don't feel normal," he said. And all of this repetitive finger movement, so important to him then, became proof to his doctors that his right arm could be saved!

Walter Reed was a challenging experience, recalls Dwyer, but he also remembers the positive influences while there, to include the doctors and nurses, and, of course, his ever-supportive family. One of his most memorable, positive, events involved his young son.

It was perhaps six weeks after he was injured, and the first time that his 3-year-old son, Tim, had been brought in to see him. They had told his son ahead of time that his Daddy had gotten hurt. That he had lost his hand, and an eye, and that he had a lot of "boo-boos" on his face, his neck, and his right arm. They explained it all the best they could to a 3-year-old. So, wife Jennie and hospital staff brought him in and put him on the bed alongside his dad.

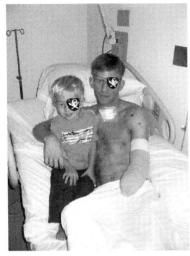

*Capt. Ken Dwyer with his son, Tim, at Walter Reed, 2006.*

"They had pulled the trach out, so I could talk a little, not much, just whisper some. So, my son looks at me and says, 'Daddy, where are your teeth?' They didn't tell him that my teeth got blown out, too. So, I told him that I'd be getting new teeth. Out of the mouth of a 3-year-old comes: 'Are they going to be white teeth?' I told him they'd be tooth colored teeth! That was the only question he had. Then he leans over to the extensive stitching on my right arm, and he gives me a big kiss on the stitch-line. Then he looks up at me and says: 'Boo-boo all better now, Daddy. Let's play baseball!'"

"So, when you think about perspective in life, and you're going through something like that (his severe wounds), all of a sudden, you're reminded of what really matters. It was absolutely a life-changing moment. And so to me, the only thing that mattered from then on, was to figure out a way to get better, to overcome the injuries that I had, so I could play again with my boy," said Dwyer.

Perhaps from the beginning of time, but more likely, much later, Moms have always solved, or at least eased the discomfort of, childhood boo-boo's, with soothing words and a kiss where it hurts. Clearly, Dwyer's son had learned that valuable 'first-aid' lesson from his Mother Jennie. And that one kiss should then be all that was needed to make Dwyer's substantial boo-boo's "all better." While it, of course, did not, regardless, at that moment, his son's gesture did give him both an emotional and a spiritual lift, and the determination to do all that he could to heal and overcome his residual injuries, so that he could return to the life he had envisioned in healthier times: Support for his family and service to his country.

A bit of combat wounds and recovery irony would occur several years later, involving another SF team member from Fort Bragg, another medic, whom Dwyer knew. This medic hadn't flown over to Afghanistan yet, and happened to be at Walter Reed when Dwyer arrived, and was among the first to see him, along with Mrs. Dwyer. Fast-forward four years, with by-then, Major Dwyer back at Fort Bragg. His medic friend, who had visited him at the medical center, had been deployed, and while in Afghanistan, he "got shot up." So now, in role reversal, Dwyer traveled up to Walter Reed, and together with that medic's wife, visited him at the medical center, and this time, it's that medic who's in the bed recuperating! Strange turn of events, and not particularly a happy time, except that the medic did survive his injuries, as, of course, had Dwyer.

After six months at Walter Reed, not the two-years originally projected by doctors, it was back to Fort Bragg and Third Special Forces Group for Dwyer (February 2007), with the idea of resuming his former duties. But was he really healed enough to actually take it all on again? "I was not! But looking back now, I wouldn't change anything. I was stubborn! To be honest, I was actually not really ready to leave the hospital (Walter Reed). I had convinced everyone there that I was. But I was still on way too many meds. And my body wasn't completely healed, wasn't completely ready."

"But I wanted to get back to my guys. My goal, while at Walter Reed, was always to get back to Bragg by the time my guys flew home from that deployment. And when they flew into the airfield there, I went out to see them come off the plane. That was pretty special, because they didn't think I was going to live. I didn't think I was going to live, either!"

While reconnecting with his team members, he learned that their firefight had not concluded in the American Afghan Army unit's favor. "After pulling out casualties and recovering equipment, the team had had to withdraw back to the Fire Base to reconsolidate. As it turned out, that whole deployment had been a bad one. And I really hated the fact that I wasn't there with them for all of it," recalled Dwyer, with a hint of sadness still, now some thirteen-years later.

In 2007, while back at Fort Bragg, Captain Ken Dwyer was presented with the Silver Star, the United States military's third highest award for combat valor, in tribute to his actions, during that 2006 ambush and intense firefight, attempting to maneuver his team members away from their vulnerable position, directing others while dangerously exposed and returning fire, against a numerically superior enemy force.

Appropriately, his team medic and his team sergeant, who were the life and death difference for Dwyer that fateful day, were also awarded the Silver Star, for their heroics under fire while providing critical medical assistance to him, as well as to others on the team, during that vulnerable ambush and battlefield fight.

While recovering at Fort Bragg, Captain Dwyer continued to actively work his way back toward full active-duty status and do so in a way that would prove personally satisfying to him. Command didn't expect him to do a whole lot, only to slowly and steadily make his way toward fulfilling the anticipated workload. One of the big hurdles in the recovery process that would eventually challenge him to the fullest was his desire to return to jump-status.

That journey began with the necessary permission to talk with his battalion commander. Said Dwyer to the commander: "Hey, sir, when can I start jumping out of airplanes again?" in classic Dwyer right-to-the-point fashion. The commander patiently reminded him that, with only one hand, he simply wouldn't be able to jump again. Dwyer's response: "Sure I can, sir. I just gotta figure it out." His next stop was to the unit's flight surgeon to see about getting a jump waiver. With the same reasoning the battalion commander had used, the 'Doc' refused to grant him the necessary waiver.

In the face of two denials, Dwyer, now more determined than ever to succeed, linked up with one of his very experienced non-commissioned officers, and the two of them spent the next three-weeks, working to figure out all of the different jump functions and factors, importantly, of course, how he could exit an aircraft safely, and then get to the ground with one good hand and his prosthetic other one. He wanted to figure out all of the contingencies, malfunctions, emergencies, etc. that could happen, and then how to deal with each. With his procedures and plans finally in place, and fully ready for that jump, he went back to see his battalion commander. Dwyer said to him: "Sir, let me show you all the great stuff I've learned." And after he had effectively and persuasively demonstrated his techniques and readiness, realizing Dwyer's firm commitment and all of his devoted preparation, the response of his commander was reassuring reinforcement to his ears. And with an unexpected twist. Said the Lieutenant-Colonel to Captain Dwyer: "Not only are you going to jump out of a perfectly good airplane today, I'm gonna be right there beside you when you do it!"

And, so, that day, he and his commander both jumped, and successfully. For Dwyer, that important step was accomplished, less than a year back from the very edge of death. "That jump was a big deal for me," he said. "To be able to go back and show people that, hey, you can do whatever you commit to doing. As long, as #1, you're creative enough to figure it out, and committed enough to do it, and #2, you have that a supportive population around you who're going to allow you to succeed." In that instance, all credit for the former went to Dwyer, and all credit for the latter to his wise and trusting battalion commander. As he later related, that particular command decision, to permit his jump, was a powerful example of leadership, especially so, in the eyes of a young captain. An example that he would long remember.

Subsequently, whenever he thinks about being a leader in the United States military, he continues to look back on that accomplishment as a lesson. "Ninety-nine-point-nine percent of the leaders in our military would have said, no way. There's no reason for you to jump out of airplanes again. You're not going to do it. It's too risky. But he (battalion commander) allowed me to be the best version of me. And sometimes I think as leaders, we get in our people's way, perhaps keeping them from peak achievement. So, instead of empowering them to be great, we tend to hold them back. I constantly look at that when I'm in a leadership position. Allowing soldiers to come up with their own ideas, executing those ideas, and developing themselves to their best potential. That's always stuck with me," said Dwyer. And a return-to-jump-status footnote: After he had successfully completed his jump, he went back to the 'Doc' and got his waiver!!

Despite his significant, life-altering injuries, as he continued to make progress and to get better, the most important issue on his mind now became keeping up with his peers. "Honestly, through that period, it was a pretty gigantic challenge to stay competitive, because I was so messed up. If you get behind, in the officer career-track of the United States military, you just don't catch up. There are just so many 'gates' that you have to complete. You have to do this type of job, you have to progress to this assignment, you have to do a broadening assignment, then you have to serve at this level, then at that level. If you miss one of those 'gates,' it's over. You simply can't catch up," said Dwyer.

That being the case, he had to work even harder than others, to stay on that officer-progression track. For about six months, then, after returning to Fort Bragg from Walter Reed, he worked in the operations section for Third Group. From there, he was picked-up to be the Third Special Forces Group Headquarters and Headquarters Company Commander. He held that responsibility for over a year, while he continued to recover. It was also during that time that he received his promotion to Major (2010).

And, at about that same time, Major Dwyer was afforded the opportunity, along with family, to attend Graduate School, under the Army's Intermediate-Level Education (ILE) Program, at the Navy's Post-Graduate School in Monterey, California. There, for the next 18-months, he studied Defense Analysis on the road to a master's degree.

Designed as a bit of a wind-down period from the demanding pace of front-line service, for Dwyer, however, it wasn't a completely satisfying time. "It drove me insane to be sitting in class knowing that I had brothers in Third Group who were back in Afghanistan, and my inability to be with them," he remembered. It so impacted his ability to concentrate in class that his professors, realizing that he was not totally focused on their lecturers, pulled him aside, early on, to ask if he was alright. Dwyer's response, said only to himself: "Look, I've got my guys who are right now over fighting a war, and I'm sitting here in a class learning the theory of insurgency. I could be over there right now actually fighting the insurgency, rather than the theory of it. I got it. Let me go fight the war."

Looking back at that academic sojourn, he admits that he did understand the value of that graduate school learning experience, much more so in the years that followed. "Really what I took away from that experience was becoming a much better critical thinker and problem solver. Not taking information at face value but having the discipline and rigor to dig deeply into something to get to the bottom, and the truth of the problem, in order to develop a better solution," said Dwyer.

And from the perspective of his family, that time away from the 'fight' gave him invaluable time to spend with his kids when they were young, along with being able to coach baseball and soccer. Doing the things he loved, with his children and other youngsters.

After his many months of re-energizing time, while away at that Naval school, with master's degree firmly in hand (despite some of his earlier feelings, he did buckle down!), Dwyer returned to Fort Bragg to work as the Chief of Readiness for Special Forces Command. After a year in that role, he returned to Third Special Forces Group to command a company (2012). The following year, Dwyer made his fourth trip to Afghanistan. And once again, he was fortunate to have leaders around him who continued to want him to be the very best he could be. They weren't going to hold him back, by assuming since he only had one hand, he couldn't do certain jobs.

"Now, they did give me a big list of things I had to do," recalled Dwyer. "In order to go back, you have to accomplish all of these things, so we know you can handle yourself, and that you can lead your men. And so I did. It was great, totally fair. Take a PT test. Got it! Go to the range, qualify with your individual weapon. Got it! Skill Level One soldier tasks. No problem. I just figured out how to do all that was required, with the use of my prosthesis, and moved on! Recalling the idea that you can do anything you want, as long as you're creative enough, and committed enough, to make it happen," he said.

Fourth tour to Afghanistan (2013). What did he remember most about that experience, back in the war zone, but unlike previous trips, dealing with the residual effects of severe wounds? "I did get a lot of funny looks from local Afghans, being an American soldier back in country, with only one hand and one eyeball! But you do need to realize that Afghans are much, much more comfortable seeing people with missing parts, than Americans are. Recall that it is the most heavily mined country in the world, even after all of the international de-mining efforts (with land mines being found that actually went all the way back to the Russians!). So, they're used to seeing people running around with missing legs or missing hands or whatever. But what they're not used to seeing is American soldiers running around Afghanistan with prosthetic arms and fake eyeballs! Honestly, to me, the reaction among the locals was funny," remembered Dwyer.

For this deployment, the nature of his job had changed. He was no longer with a Team out there running missions every day. He was occupied with command-and-control responsibilities, far more so than with his previous times in country. "I did go out on a couple of different small level ops, but nothing like those from past missions," he remembered.

After this six-month deployment as a company commander, he returned to Fort Bragg. Next, leaving operational Third Group, he went on to take another company command at Bragg's Special Warfare Center (SWC), the Special Forces 'schoolhouse.' He took command of the company that trains '18-Alphas.' "When a Regular Army soldier wants to become a Special Forces guy, he goes to Selection. After that, he goes to the Q-Course (Qualification). During the Q-Course, he has a four-month period where he's trained in his specialty. So, all SF officers, during the years 2013-2014, who went through the Q-Course, spent four months with me, before they were able to move on to the next phase," said Dwyer. He instructed these select combat leaders on the planning principles specific to Special Forces missions, along with other pertinent subjects, relying heavily on his own experiences in combat.

Major Dwyer truly enjoyed his time there instructing future Special Forces officers. "That was a tremendous job. It was the second-best job I've had in the Army, second only to being an SF Team Leader. Having the opportunity to significantly impact those young Captains' lives and careers on a daily basis," he said.

As such, he poured heart and soul into his class preparations, and explanations for his students, to include both the experiential highs, and the not-so-highs, of his time in the Army. "What I brought to the task each day was an effort to make them better than I was. So I took all the mistakes that I made, not only as a Team Leader, but in the other aspects of my career, and I packaged them up, sat the students down, and talked to them about all the things that Dwyer screwed up, so, hopefully, they would perform better," he said. As can readily be seen from his career, marked by exceptional courage, while he may have messed up occasionally, like every other soldier and civilian on earth, his successes, his real-world actions, the performance positives that he was able to impart to those young officers, were far more impactful and lasting, than any classroom lecture or power-point could ever be.

In mid-2014, Dwyer moved up to be the Executive Officer of his SF Group. Then in late 2015, Major Dwyer became Lieutenant-Colonel Dwyer, worthy beyond question of that important promotion to senior officer status. With that, he went back to Special Forces Command to become the Chief of Plans & Operations (2016-2017).

*LTC Ken and Jenny Dwyer
with son Tim,
and daughter, Julia,
in-residence,
Hunter Army Airfield,
June 2017.*

To bring his career assignment progression up to the present, in the summer of 2017, LTC Dwyer made the move to Savannah, Georgia for a two-year tour as the Garrison Commander for Hunter Army Airfield. In the collective opinion of his Garrison colleagues, the commanders of the multi-branch military tenant units on Hunter, and the feelings of the many community leaders whom he has touched through his many public speeches and his winning personality, LTC Ken Dwyer has done an exceptional job with his duties and responsibilities, both on post and out in the community where he continues to make friends and impress people every day. Ken Dwyer has been a truly effective and impressive ambassador for both Hunter Army Airfield and the United States Army. We are collectively sad to see his tour with us come to an end, which it will on June 13, 2019.

*LTC Ken Dwyer,
Garrison Commander,
Hunter Army Airfield,
2017 – 2019.*

After some well-deserved family vacation time, LTC Dwyer will be returning to Fort Bragg where he will be assigned to the U.S. Army's John F. Kennedy Special Warfare Center and School as the G-3 (Army designation for the Operations Officer serving on the General Officer's staff).

Following the career progression, significant combat injuries, and road to recovery portions of this narrative, the Colonel was asked to share his feelings about some of the basic values and expectations that he sees now, and has experienced, as a senior Army officer through his 20-years to-date of outstanding service to our nation.

## PERSPECTIVES ON LEADERSHIP

*How do you define Courage?*

"Being able to overcome fear. With the things we do every day in combat, you're going to be afraid. But having the ability to get past that fear, and do what's needed and right, regardless. That, to me, is courage."

*Looking back over your Army career, what do you consider to be the most significant life-lesson that you've learned?*

"Don't ever quit, no matter how bad it gets, no matter how hard it seems. That's one of the things that I hope my kids, my soldiers, and people who are around me, can also take away from my situation. That it's not OK to quit doing what you love, or what is the right, or is the expected thing for you to do, just because it's hard."

*What do you believe is the most important element of a leader's character?*

"Selflessness. To be more concerned with the well-being of others than that of yourself."

*Can you provide a memorable example, that you've observed, of exceptional character?*

"The first two people who come to mind, whenever I would think of anything in that context, would be my medic and my team sergeant. They're the two best men that I've ever met in my entire life. That day, when I got hurt, they potentially sacrificed their own lives to save mine. And, wow, there's no greater gift than that, right? They put their own lives on the line. And they didn't just do it that day, they did it every single day. I keep this picture here (enlarged photo displayed in his Garrison office) because those are my guys. And the most important things in your (combat) life are those people on your left and your right. And it's that brotherhood, where every guy in that picture would've given his life for every other guy. This could also be an example of courage as a component of character."

*Do you think leaders are born or made?*

"I think it's a combination of the two. I probably lean toward greater than fifty percent 'made.' The experiences that you go through in your life make you the person that you are. Every leader has to be given certain qualities, have a certain raw material within him or herself, that can be fashioned into the leader they seek to become. Something that people don't like to think about or hear, is when you talk about fashioning character, or talk about creating leaders, it's the pain and suffering in our lives that creates greatness. It's those cold, wet, tired, miserable moments at Ranger School. It's the I can't breathe, but I still have to accomplish my mission, in Dive School. It's the people are shooting at me, but I still have to do my job. Moments in life that may hurt, but that end up making us great people."

*In your view what makes, and keeps, a leader effective?*

"He's got to focus on why he does what he does. Everything I do in my military career is focused on making sure my soldiers can accomplish their mission. So, you've got to have a 'why' in your life, and you've got to understand your 'why,' and you've got to commit to it every day. Whenever I talk to people about leadership, I always come back to <u>the</u> most important thing that a leader can do, or quality that he can have, which is that he must be more concerned with the people he leads, than he is with himself. He must care more about <u>their</u> success than he is with his own."

*Is there a memorable example that you've observed of exceptional leadership?*

"Absolutely! My battalion commander when I got hurt. He was awesome, he was unique. He was very passionate about his job. Now, I'm going to be in the minority within the military when I say this, but we have a whole lot of leaders who don't understand the value of emotional leadership. They understand the science of it, they understand numbers and facts, and computing things and cost-benefit-analysis. They understand all these things very well, much better than I ever will, but I think we have a significant problem with our leaders not understanding the emotional value of leadership. Through words and actions, making it evident that a leader truly cares for his soldiers. My battalion commander, back then, understood that in spades."

*What have you most valued about your military career?*

"What I most value is the brotherhood. It's being part of an organization that's bigger than yourself. That's what I value the most about the Army, and what I'm going to miss the most when I walk away someday. Not being a part of a team anymore."

*What has your military service meant to you, and given back to you?*

"I will never be able to re-pay the Army for what it's done for me as a person. And I think that anybody who comes into the military, and throws their whole heart into it, will absolutely tell you that serving has made them a better person. People look at me like I'm crazy when I say that, just because of my situation with my injuries, but I absolutely do believe that I'm a better person than I would have been, had I chosen not to follow this path."

Service in the Army started out as something I wanted to do, as a short-term thing. But honestly, once I really got into it, I fell in love with the teamwork, and the comradery that exists, and decided to make it a career. Perhaps the key attraction, the key component, is the common purpose everyone has. I just love being around soldiers, every single part of it, even the bad days with young soldiers when they're making mistakes and messing stuff up. It's still an opportunity to give them a chance to grow and get better. It's an unbelievably rewarding job. I can't imagine a life where I would've done something different, and not having the sense of purpose that I have every day with what I do."

*Colonel Ken Dwyer*

**UPDATE**: Following his command at Hunter Army Airfield in Savannah, LTC Dwyer was assigned as the Special Operations Center of Excellence Operations Officer at Fort Bragg (Liberty), NC (2019-2020). He was then selected to attend the Army War College (AWC) in Carlisle, PA (2020-2021). Following graduation from the AWC, he served as the Deputy Commanding Officer of the 7th Special Forces Group at Eglin Air Force Base in Florida (2021-2023). While at Eglin, LTC Dwyer was promoted to Colonel in 2022. Currently Colonel Dwyer serves as the Commander of the TRADOC (Training and Doctrine Command) Leader Training Brigade at Fort Jackson, in Colombia, South Carolina.

# CHAPTER 5

## "That Was the Most Turbulent Rescue
## I've Ever Done!"
## United States Coast Guard Air Station Savannah
## Captain Brian C. Erickson

*Captain Brian C. Erickson,
Commanding Officer,
Air Station Savannah,
United States Coast Guard.*

From 2018-2020, Captain (Select) Brian Erickson served as the Commanding Officer of Coast Guard Air Station Savannah (GA). A native of Port Townsend, Washington, Brian Erickson is a 1998 engineering graduate of the United States Coast Guard Academy. He would later go on to earn a Master of Science degree in Aeronautics and Astronautics from Purdue University. His pathway to Coast Guard aviation, and eventually Coast Guard command, was initially, as you'll see, neither obvious nor predestined.

"When I was in high school," remembered Erickson, "all I really wanted to do was ride my dirt bike and drive around in my truck. Needless to say, I didn't get very good grades in school." But as can happen, less-than-great-grades in school isn't always a true indicator of one's innate intelligence, of one's 'smarts,' if you will. And such would clearly prove to be the case with Brian Erickson, who would go on to excel in a very demanding and challenging aviation and command career.

But back to those 'average' grades in high school. A couple of his buddies were talking about joining the Coast Guard after their senior year. Erickson well remembered that he very much <u>did</u> want to leave his hometown, but definitely did <u>not</u> want to go overseas, "so the Coast Guard sounded like a pretty good fit." Impacting his service choice, as well, was the fact that the Coast Guard Cutter POINT BENNETT (with a 10-person unit) was moored right there in the Port Townsend marina; then during his senior year, he spent one assigned outreach period each school day with that unit; and, finally, a retiree from the Coast Guard, whom he'd known for many years, managed one of his dad's businesses. No question, he had grown up amidst some very prominent and positive Coast Guard influences. So, in 1992, he made the decision to join, but would need the help of his parents to sign him into the military, since he was just 17 at the time. Then things happened quickly. One month after high school graduation, Seaman Recruit Erickson was off to Coast Guard boot camp at Cape May, New Jersey.

From Cape May, his first assignment was to Key West (FL), where, with the rank of Seaman (non-rated sailor), he would be aboard the USCG Cutter SEA HAWK, a unique high-powered, solid-hull, hovercraft (called a "surface effects ship") being utilized for law enforcement, counter-narcotics, and alien migrant interdiction operations in the Florida Straits. During his one-year serving in Key West, he decided to put his name on the list for more training, applying, then, for aviation electronics technician school. "My thinking was that I'd just go ahead and do that for the next four years. Frankly, at that point, I really didn't know what I was going to do long-term, you know, career-wise," said Erickson.

But then, on the subject of career, opportunity unexpectedly knocked. He received a letter stating that his test scores were high enough to make him a candidate, not just for aviation tech school, but potentially high enough for actual Coast Guard Academy eligibility! Erickson called his mom to ask if she thought he should apply to the Academy. Without hesitation, she responded: "Oh, my God, yes, you should." That left little question in his mind that she was, indeed, supportive!

So, with Mom's seal of approval, he applied to enter the Academy. He was accepted to attend, first (at that time), the Naval Academy's Prep School for one year of preparatory study. Back then, recalled Erickson, the Navy's prep school accepted prior-enlisted from the Marine Corps, the Merchant Marine, the Coast Guard, the Navy, of course, and also top graduates with Congressional appointments coming directly from high school. Following his very good year at Navy Prep, as expected, Erickson moved on up to the Coast Guard Academy proper (New London, Connecticut). Year One on the road to becoming a Coast Guard Officer. Best yet, upon entry, with his enlisted time, he already had "2-years and 5-days" of service time on the record!

Cadet Erickson attended the Academy from 1994 to 1998. For co-curricular activities, he played lacrosse (club sport at that time), along with several other intercompany sports. He majored in naval architecture and marine engineering (single degree program). He served as class vice-president, and later president, which were definitely honors, and a clear indication of his stature among his Academy peers.

Ensign Erickson's first assignment was aboard Coast Guard Cutter MELLON, a 378-foot-high endurance craft, based in Seattle, Washington. He served there, "on the engineering side of the house," as both an engineering-officer-in-training, and as a trained damage control assistant. He would go on, in short order, to qualify as a deck watch officer. Then, just three months after his arrival on the Cutter, he applied for flight school and following a lengthy process and wait time, he was accepted. Shortly thereafter, he "left the boat." Total time with Cutter MELLON: One-and-a-half years.

At the start of his Coast Guard tenure, with all manner of career options on his mind (in or out of the CG), he recalls having had no real expectation of one day going to flight school. "It was something I didn't even think I could do," said Erickson. Back in Port Townsend, "my father was a huge general aviation influence (obtained a private pilot's license & owned his own plane), and I grew up with so much enjoyment around flying. I eventually got to the point where I did want to make it a career, so that 'I wouldn't have to work a day in my life!" As if by fate, somehow, amidst the rigors and demands of the Academy, while there, he actually found the time to obtain his private pilot's license. As an Academy graduate with an impressive all-around record (not to mention already having 150-hours of civilian flight time!), LTJG Erickson was accepted for school, and began his flight training in November 1999, at the Navy's flight school in Pensacola, Florida. This is the training site for all future Navy, Coast Guard, and Marine Corps pilots. Adding-in occasional joint training exercises, Air Force pilots became involved there as well. Rounding out the student mix, select pilot-trainees, from approved allied nations, also come to Pensacola for flight instruction.

Following flight school graduation, LT Erickson's first duty station as a Coast Guard helicopter pilot was Air Station Port Angeles, Washington (February 2001). Convenient to his hometown, it's also where he first met fellow CG pilot Marshall Branch. He recalls flying a rescue mission with him. The memory and significance of that first meeting is that the now-Captain Marshall Branch would actually end up immediately preceding Brian Erickson as Commanding Officer of Coast Guard Air Station Savannah.

Later that year, September 11, 2001, to be exact, fanatical Islamic terrorists carried out their destructive attack on America, causing more deaths, in three intended aircraft crashes, than resulted from the equally unanticipated surprise attack on America by Japanese forces at Pearl Harbor, almost 60-years earlier. Missions were immediately altered for America's military everywhere, to include Coast Guard personnel at Air Station Port Angeles. "I remember driving in to work on September 11th, and over the next two to three weeks, operations definitely changed, as you'd expect," recalled Erickson. "We were much more focused on the port and surrounding waterways, patrolling the local areas by air to ensure that the area waterways and infrastructure, were secure."

It was close to that timeframe when LT Erickson took part in his very first rescue mission, launching from Port Angeles. Serving as co-pilot, their flight was at night, in fact, a very dark night, with low visibility across a turbulent Pacific Ocean. They were searching for a man who had wrecked his boat on some of the very treacherous rocks, somewhere out along the coast. With the high waves, his boat was crashing and breaking up against those rocks. The crew was told that the man was able to get out of his boat and had made it to a place called Little Rock Island, an outcrop located about 500-feet from the actual shoreline. With Erickson's helicopter now about 70-miles from launch point, they finally saw a flare in the darkened night. With that sighting, they approached the location, got into a low hover, and could finally see the man, who wasn't standing, but out of the water, lying down, with the waves crashing near him.

"So, we put the rescue swimmer down, who determined quickly that the man was having heart attack or stroke-like symptoms. We really couldn't see anything except where our light was pointed directly down on the man and our swimmer as he did his work. Once stabilized there on the rocks," said Erickson, "we hoisted them up, and in the midst of increasingly high winds, dark skies, and low fog, we then navigated our way back to Port Angeles, and directly to the hospital, where the man arrived alive and stable." Erickson indicated that, three days before, he had flown a previous search mission, out over the water, looking for a father and his two kids who'd gone out in the fog, but they were never able to find them. So, sadly, no rescue. The bodies of the missing three were located weeks later. The Little Rock Island flight had, thus, been his very first hoist-and-rescue mission.

It was 2003, while still at Port Angeles, when he began flying his first missions as Flying Pilot. Many of these were medivac flights. There was a private airlift service in the area, but once the weather got bad enough, they wouldn't fly. So, it was left to the Coast Guard to make the medivac flights, regardless. The routine would be to fly out to the Indian reservations at Neah Bay, located at the very tip of Washington State, land on a helicopter pad at the Coast Guard Station there, to pick up an individual who needed to be moved inland to a higher level of care. Most often at night, it seemed, navigating through fog, depending totally on radar, "creeping through a very treacherous environment" to get out to that Station, and then back. There were a number of other missions flown while Erickson was still functioning as flight pilot, to include "hoisting a guy off a tanker ship in the daytime, so that wasn't too bad," and then actually landing on a Navy ship at night to take a person off, and then fly on to the hospital for treatment.

Then began his rescue missions as a Coast Guard Aircraft Commander, the first one occurring in 2004, while at Port Angeles. At that time, LT Erickson and his colleagues were covering for Air Station Astoria, which resulted in a much larger area of responsibility, taking them all the way down to the mouth of the Columbia River. Flying in an area where they didn't normally operate, they would be looking for a crew member out on a fishing boat, thought to be possibly suffering from pancreatitis, said the Flight Surgeon prior to their leaving Port Angeles. As Erickson's team headed down-state along the coast, it was, as he recalled, "a beautiful night, as we flew over the Olympic Mountains with a clear night sky." They then dropped down on the southwest side of Washington State and headed out to the fishing fleet and found the vessel in question. At about the time his helicopter had initiated a circle around the boat, they got a radio call from the Flight Surgeon updating the crewman's prognosis. He did need to be medevac'd.

This fishing boat was a traditional one, with a lot of rigging and lines, meaning that the rescue basket would need to be carefully threaded down through that maze. With the horizon now dark, Erickson's night vision goggles were actually more impediment than help. So, he flipped them up to get away from having "two little TV screens that were just giving me green lights." But then, the challenge became his point of reference, since he was looking down at the boat, now with its deck in ocean-water motion, making the precision insertion required all the tougher. After sending a trail line down to the awaiting boat crew below, with great care, Erickson was able to steady his aircraft so that, despite high seas below, he was able to thread the rescue basket ever-so-carefully through the cables and rigging on down to the deck of that seas-shifting boat. With the trail line to assist, boat crewmen were able to pull the basket on in. Due to the added danger of entanglement, no rescue swimmer was sent down, or even needed on this one, since radio contact was made and maintained with an assisting boat crew member on board. Thankfully, with the patient actually able to walk, and with crew help, he was able to position himself in the basket. Once safely strapped in, he was hoisted carefully on up to the aircraft without incident. Erickson flew him to a waiting ambulance on shore. The man was hospitalized in stable condition, and he survived. Many additional search and rescue missions followed during his time at Port Angeles. Along with his Aircraft Commander designation, Erickson also served there as Flight & Ground Safety Officer. Also while at Port Angeles, he deployed aboard a 210-foot CG Cutter for 45-days, serving off the coast of Mexico.

Next assignment: The Coast Guard's Polar Operations Division, from 2005-2006. Providing him with the parting thrill of landing on top-of-the-world ice, the Coast Guard was, at that time, actually shutting down its aviation polar operations division. During this three-month tour, Erickson, serving as the aeronautical engineer, had the opportunity to take the last Polar Operations Division trip to the North Pole! And that was only the third time a U.S. surface vessel had ever reached the Pole. Their ship was the HEALY, a 420-foot Coast Guard Cutter ice breaker, with two of Erickson's helicopters on board. The mission was a scientific one, with the aviation detachment supporting the scientists on board. Erickson's group then consisted of four pilots and four mechanics. Following their work at the Pole, "we came down the backside and went to Norway and Ireland, before the ship eventually made its way back to Seattle," he remembered. While Erickson was operating in the remote North Pole, his wife Leighanne was 'holding down the fort' in Mobile, Alabama, as Hurricane KATRINA made landfall only 65-miles to the west and knocked down trees throughout the Mobile neighborhood where they lived. Thankfully, none fell on their house! Erickson's North Pole assignment was four-months in duration.

From 2006-2007, LT Erickson served as the Assistant Aeronautical Engineering Officer at the Aviation Training Center in Mobile (AL), providing him with a world apart climate and temperature change! That came about because Mobile was, at that time, the headquarters for the Polar Operations Division, which, while he served there, the Coast Guard was in the process of dissolving, as previously mentioned. As the Engineering Officer, he was in charge of all of the H-65 "Dolphin" helicopters at the Mobile Center. "I oversaw the mechanics, all of the maintenance, and all of the warrant officers. Everyone who turned a wrench on the H-65's became my responsibility, in order to ensure the air worthiness of those aircraft," said Erickson.

Then came a major change from the normal assignments. LCDR Erickson was competitively selected by the Coast Guard to spend two years at Purdue University (2007-2009) to study for, and earn, a master's degree in aerospace engineering structures. Thinking back, when he entered the program there, still earning his normal Coast Guard salary, he had already accumulated 15-years of service (Enlisted time + Academy years + Officer active duty), making him, at his level of compensation, no doubt, "the highest paid master's student there!" Erickson recalled that he was being both well-paid and with "no responsibility!" The author would respectfully step in and disagree. In a demanding advanced degree engineering program, at a highly regarded engineering-oriented university, and with the Coast Guard no doubt keeping an eye on its "investment," there is no question that Erickson had ample expectation and "responsibility." No doubt the reason he felt that way, in hindsight, was the fact that he was an exceptionally bright, capable student, studying in a field in which he excelled (as with every Coast Guard assignment throughout his career). That combination of attributes and abilities can certainly take the edge off anyone's course work and degree effort!

Following his graduation from Purdue, LCDR Erickson began a three-year tour at Aviation Logistics Center (ALC) Elizabeth City (North Carolina), where he was assigned as the C-130-H Product Line Engineer. He had done fixed-wing piloting in the primary course at flight school, prior to moving to helicopters in advanced, and then he completed a ten-week C-130 flight course in Clearwater (Florida). He was responsible for the maintenance and logistical support of the entire C-130-H fleet there. At the time, that fleet consisted of 24 aircraft globally deployed. "I loved that job. It was so much fun flying that plane, and I really enjoyed the work," he remembered. Principal reason why? "We were making the aircraft better for years to come. Anything that was prototyped on the aircraft, like a new GPS, night vision goggles, or a moving map for the pilots, or a safe to store classified material. We also installed the first center wing box in any Coast Guard C-130, which essentially extended their life for another thirty-years." said Erickson.

At Elizabeth City, he was the engineer responsible for signing off on the approval for any flight following maintenance or repair, and that responsibility carried over to anytime an aircraft was damaged. As an example of the latter, they had a C-130 out in the Pacific with a load of U.S. Congressmen on board, who were visiting a Pacific Island. On departure from the island, the plane hit a bird, forcing it to circle around and land back on that island, leaving passengers and crew stuck there. In that particular instance, "they contacted me by cell phone, and we immediately developed a suitable repair plan to be accomplished right there, in order to make the aircraft once again safe for flight. Challenges and solutions like that made it very enjoyable work for me," said Erickson.

The remaining question, of course, was what about the Congressmen? "Fortunately, there was enough room on the island to land a second bird. So, we were able to load and fly them out, while the repair party remained to fix the old aircraft, and we were eventually able to fly it back," he said. They also did logistics missions during his time at Elizabeth City. In one such, Erickson piloted a C-130 down to Guantanamo Bay, flying "right over the top of Cuba." Later, aerial observation after the Haiti earthquake. And several flights over the "Deepwater Horizon" major oil spill, caused by the explosion of the large oil drilling platform there, seeking to determine the extent of the environmental damage from that huge oil float.

Then, in 2012, LCDR Erickson began his first tour (3-years) at Coast Guard Air Station Savannah Ga. He served as the Aeronautical Engineering Officer, the junior position among the four officers on the Air Station command staff. That assignment allowed him to both stand-duty and fly H-65 rescue missions. One rescue he flew on involved a fishing boat that had run aground at night, in very bad weather, near Jekyll Island Ga. Although he was the Aircraft Commander for this particular flight, Erickson, to enhance training opportunities, elected to put a junior officer in the pilot's seat for this mission and "he did a terrific job," recalled Erickson. They launched and "navigated down through some pretty hairy weather." After searching, they found the boat that had made the rescue call. Using night vision goggles, they could see that the water was crashing over the back of it, breaking the boat up before their eyes.

Meanwhile the helicopter, too, was being "pummeled by gusty winds, oddly gusty winds," which was very unusual, he said, since the rescue was occurring in the middle of the night. "We really had tried to hoist the swimmer down to the boat, but it continued to be just too tight, trying to work his way through all of that rigging, even though, strangely, at his point, the boat had actually become relatively stable," he recalled. Erickson remembered that the boat's crewmen really wanted to get off that boat, as they watched it breaking up under them, even though, at this point, they were likely only in five or six feet of water. "They were within 100-feet from shore, but completely unwilling to venture out themselves," remembered Erickson. So, to avoid rigging entanglement entirely, they ended up sending the rescue swimmer down directly into the water, had the crewmen jump into the water, and then the swimmer pulled them clear of any potential danger from their wrecked boat. Erickson's crew hoisted the crew up, one at a time, and flew them over to the nearby airport. While that delivery was going on, given the time required to arrive on scene, and the time it took for the rescue, his helicopter was running low on fuel. So, they left the rescue swimmer in the water, letting him swim to the beach. Now that might sound strange, but it's what these amazing rescue guys do, they're swimmers(!), and they maintain peak physical conditioning, due to the rigors of their job. Erickson then came back to pick up his swimmer, then headed to St. Simon's Island to refuel, before flying back to Savannah.

One of his most memorable rescues occurred in the daytime, "memorable for among other things because it featured "some of the highest seas I'd ever encountered." It involved sailing vessel "Molly Jane." A call had come in that the crew of this vessel had managed to ride out the night on board and were now located 41-nautical-miles southeast of Charleston (SC). They were caught out in Tropical Storm Andrea (June 7, 2013), so by this time, they had lost propulsion, their boat was flooding, and the main sail had ripped! With their boat dead in the water, they wanted off! A C-130 flew out and found the boat. Pin-pointing the exact location, the Coast Guard then sent out an 87-foot Cutter to either tow the vessel in or at least get its crew on board. But the waves were too high, rendering it "out of limits" (i.e., sea-state too high), so they couldn't safely make the surface transfer. So, the next rescue call came to Savannah, and two helicopters were sent. Normally, one would have already been at Air Facility Charleston (serviced by Savannah), but the daily crew rotation was underway, with that crew now back at the Air Station, so there was actually no helicopter available at that time in Charleston.

Erickson was piloting one of the two Savannah helicopters. "We had flown a long way to get to the scene. We did have a tail wind going up, so we knew we'd have a headwind coming back, meaning we had to be very careful with our fuel planning," he recalled. It was about 2 PM and he remembers that it was tropical- storm-windy, and with it, the seas were quite rough ("17-foot swells at the buoy"). "Even out in the Pacific, doing work from Port Angeles, I hadn't seen seas like that very often," he said. Given the dangerous conditions, not only for the boat crew, but also for Erickson's crew, he came in at a lower altitude, safely off from the sailing vessel. "We don't normally hoist below 30-feet, but I wanted to be just above the waves for this one, so we were probably at just about 20-to-30-feet," he recalled.

He lowered the rescue swimmer, and then had two of the boat's crew members jump into the water, where the swimmer then held onto them, and helped each one into the basket for the hoist. Erickson recalls that his swimmer was tiring from the beating of the waves and the time it took to corral the basket and hoist each one up. The second CG helicopter, piloted by a lesser experienced crew, chose to hover much higher for the rescue than Erickson (about 80-feet), taking more time in- between hoists, and causing the swimmer to blow much farther from the vessel between hoists. This added time and complexity to the rescue of the first two crewmen, a fact that Erickson would use for after-flight review training back in Savannah, later on. He had comparative video of the rescues to use in illustrating his subsequent instruction, since that original C-130 had remained on the scene videotaping the dual rescue! Tired and understandably frightened, thanks to Coast Guard Air, all four boat crew members involved in that rescue survived their ordeal in good health.

Following his initial three-years in Savannah, LCDR Erickson's next assignment (2015-2018) took him to Coast Guard HQ in Washington, D.C., for a staff assignment in the Office of Budgets and Programs. His primary responsibility was to assist in annual budget development for the entire Coast Guard, along with helping to prepare their top HQ commanders to testify before Congress, working hand-in-hand with their Congressional Affairs staff. "Basically, we helped propel the major Coast Guard strategies forward," said Erickson. And it should be noted that his position at CG HQ was a special staff assignment, meaning that he had to interview for it in D.C., and was ultimately competitively selected.

After his time in D.C., it was a return to both an operational assignment and to Savannah (GA), where, on this tour, now-CDR Erickson would take command of the Coast Guard's Air Station there (2018-2020). Air Station Savannah has 5 H-65 helicopters and 110 uniformed personnel. The demanding rescue flights would begin, again, a source of both challenge and pleasure to the commander, since flying had clearly become his first love (professionally, and with his civilian license & plane), with the principal mission of assisting those in distress, which is, after all, what he'd signed up for those many years ago, when he raised his right hand and joined America's Coast Guard.

When first flying again, after any staff assignment, even an experienced pilot like Erickson is required to take a bit of a 'back-seat' for a while. So he would fly the next few missions as the co-pilot "because when you come back as CO (Commanding Officer) from a staff slot, it takes about six flights before you re-qualify for aircraft commander status," he remembered. Proper protocol, then, is typically to take some time after arrival, perhaps six-months or so, do some flying in the interim, before beginning that official six-flight aircraft commander sequence. And there's a reason. Experience had taught him that jumping to the head of the line over other pilots, who had been waiting their turn, never did sit well with the unit, which is not the best way to begin a command!

And one of those co-pilot missions would turn out to be "a really hairy one." It involved a rescue from the back of a cruise ship. Flying with LT Crystal Barnett as pilot and aircraft commander, he remembered that, when they departed Air Station Savannah, the weather was fine. But by the time they reached that cruise ship, some 35-miles off the coast, the weather had turned bad. The wind, as he described it, was "horrible." Adding to the wind and the rain issues, this rescue took place in the dark, at midnight. "It was overcast, at about 300-feet, so we didn't have any visibility up above, because of the rain. I estimated seven-foot seas and winds at 35-knots," he recalled. So, to review the mission scene they encountered: high winds, high seas, low, overcast skies with rain, a cruise ship rescue, and it's midnight. Far from an ideal mission scenario, without question, an incredible understatement!

The person in need of rescue, and medical assistance, was an elderly female passenger, who had fallen down some stairs on the ship, resulting in "multiple traumas." Erickson and crew were battling to hold an uneasy hover over the back of the ship, while the rescue swimmer was lowered to the rear pool deck on the first descending hoist. On the second down hoist, a litter was lowered for the injured passenger. Rather than using a basket in this case, the litter is a survivor-gear option that, when the situation requires, allows the patient to be laid out flat. Erickson would later label this night as "the most turbulent rescue I'd ever done." "There we were, getting beat up by the wind. But it was so dark, that we felt it would be less dangerous to just stay where we were, despite the punishing winds, than to initiate some sort of an orbit, either in the clouds or below them, and wait for the rescue swimmer to 'package' the survivor," concluded Erickson.

So, they remained, right where they were, maintaining that challenging hover, as they waited for the preparations down on deck to be completed. "It took 30-minutes for the swimmer and the ship's doctor (whom we did take back with us) to stabilize the patient on the litter, ready for hoist," said Erickson. All during that time, he recalled the sensation of almost flying sideways, as the wind was coming at them over the top of the cruise ship, while LT Barnett, a very experienced pilot, struggled to keep their hover steady. And the rain had now become blinding. "At one point, she asked the flight mechanic to reach out and wipe the rain bubbles off her side window, because with all that rain, the light coming up from the ship was distorting the view, making it difficult to see clearly, and hampering her ability to maintain our hover," he said.

With preparations finally completed on the deck below, having taken much longer than anticipated, the first hoist up would be the patient, shielded from the rain and safely strapped onto the litter. Normally, then, the next hoist up would be, first, in this case, the ship's doctor, followed by the rescue swimmer. But given unanticipated additional time on station, in order to conserve both time and fuel, the decision was made to send the hoist "hook" down and bring both the doctor and the swimmer together in a double pick-up. The doctor and the medically trained swimmer would then both tend to the injured woman during the flight to shore.

And that flight would still not be without its challenges. Heading off, then, into the darkness, with clear visibility a continuing issue, they opted for a manual departure "to avoid having the autopilot take us into the sea." At that point, "all of our focus was on the gauges, because since we couldn't see anything outside, we wanted to be sure that we were, indeed, climbing away from the water!" he said. Once they had ascended above the cloud layer, at about 1,000-feet, they brought on some autopilot and continued their climb up to 2,500-feet. They were at last out of the rain squall that had surrounded the ship and were now turned back toward shore. With the weather now much better, they could see in the distance, the welcoming lights of Savannah! They flew those 35-miles directly back, landing on the mid-town Savannah helicopter pad at Memorial Health University Hospital. With the expert care received in route, fortunately for all, the patient arrived there in stable condition. The team at Memorial made it clear to the Air Station Savannah crew, that "she had survived because of us," remembered Erickson, thankful that all involved had made it back safely.

And 'co-pilot' Commander Erickson vividly recalled that LT Barnett had done a tremendous job flying this rescue, under very trying and hazardous conditions. "That was the most turbulent rescue I've ever done," he recalled. Since he was then commanding the unit and, as such, hadn't flown nearly as many missions as he had during his first Savannah tour, after that very vivid, tense, and demanding rescue, for the only time during his command, he recalled thinking to himself: "I'm too old for this stuff!" Now, in the interest of interview accuracy, it should be noted that the Commander's actual word there was not "stuff"!

*Hurricane DORIAN Operations – CDR Brian Erickson briefs Mayor John Tecklenburg of Charleston, South Carolina prior to infrastructure overflight following the devastating impacts of Hurricane Dorian.*

During the 2019 hurricane season, Hurricane DORIAN (September 6, 2019) destroyed areas of the Bahamas, but then steered offshore as it approached the Georgia and South Carolina coasts. The Air Station Savannah teams were poised and ready to respond, but DORIAN's path skirted the coast which, while a relief to area residents, "was kind of a letdown" for Erickson's new crews anxious to put their hard training to work. But just days following DORIAN's movement away from the area would come the one heroic mission, performed by members of his Savannah team, that Erikson considers to be the "landmark rescue" of his command assignment. The M/V GOLDEN RAY was a fairly new (2017), 656-foot-long cargo ship, designed specifically as a vehicle carrier. Loaded with over 4,000 new Hyundai and Kia cars, GOLDEN RAY departed the Port of Brunswick, Georgia, on the night of September 8, 2019, and shortly thereafter capsized, ending up lying on its side in comparatively shallow water (estimated depth: 31-feet). Air Station Savannah got the call, and responded with two helicopter crews, flying at night, for the trip to Brunswick, where they would focus on the anticipated rescue of the ship's crew (joined throughout the operation by CG boats, an important assist to the successful rescues). Thanks to the rapid response by that Savannah Coast Guard team, along with other rescue personnel arriving later, there were no fatalities among the 24 ship's crew on board, when there easily could have been, since there was a fire burning in the rear of the ship, and several crew members remained trapped below for over a day. They were ultimately rescued, alive, only after holes were cut through the ship's hull by land-based rescue teams with heavy equipment!

All of the responding Savannah Air Station crew members performed heroically, and, perhaps most exceptionally, Coast Guard Rescue Swimmer Nate Newberg who, in the darkness, was lowered down from his hovering helicopter, stepping off onto the pilot house wall of the severely-listing ship, and with great courage and ingenuity, was able to rescue the ship's captain from the bridge via hoist up to the helicopter, and then by somehow finding and securing a fire hose to the pilot house wall, he was able to also rescue the Brunswick-based river pilot, trapped by debris at the other side of the bridge, by helping him climb out and then guide him down that hose to the water line, where he was then able to step right onto a waiting boat, which had arrived there from the Brunswick Coast Guard Station. Intense rescues completed; Newberg would eventually be hoisted back up to his helicopter.

For his skill, bravery, and ingenuity, operating under incredibly challenging conditions, Nate Newberg was subsequently awarded a meritorious promotion to AST-1 (Aviation Survival Technician First Class), and was also presented with an Air Medal. Ultimately, he was invited to Capitol Hill in Washington, D.C. where he was honored as the 2019 National USO Coast Guard Service Member of the Year, in the company of the Coast Guard Commandant, Congressman Buddy Carter (Georgia-1), and other dignitaries.

Eric Young, the rescue swimmer from the second Air Station Savannah helicopter, was lowered down to another section of the ship, and was able to rescue three crew members, all by hoisting. Executing challenging rescues, as well, he was awarded the Air Medal and was named the 2019 Air Station Savannah Coast Guard Person of the Year.

*M/V GOLDEN RAY Rescue Crew – CDR Brian Erickson, Commanding Officer of Air Station Savannah stands with his aircrews who responded to the capsizing of the M/V GOLDEN RAY.*

Commander Erickson remained ashore, serving with his operations team, as "quarterback," managing response risk, while his on-scene teams down in Brunswick were achieving those amazing rescues. "Long before this or any rescue, as the CO, you try to set up a culture at your unit, so that you have rescue teams that are willing to make on-scene decisions, willing to take warranted risks, for the saving of lives and property. I'm very proud that, on that night, they took every risk necessary to save those lives. And they knew that I would have their back. Savannah is a forward leaning unit. We wanted to get those people off that vessel, and we went to great lengths to do that. We will use fire hoses to rescue, we'll land on the side of that overturned ship (which one of his helicopters actually did!), even though not approved. We'll take that risk for the saving of life. That culture, that trust. You may not see that in every unit," he said.

Commander Erickson was then asked what the best memories of his Savannah tour might be, due to wrap-up soon. "I'll most remember all of the success that my people have had. My most enjoyable moments are pinning on an enlisted promotion and signing my name to a promotion certificate. That has been more enjoyable than I would ever have imagined," he said. Then, referring back to the heroics of the GOLDEN RAY and other meritorious rescues, he spoke of the enjoyment and pride he felt in "parading my people around who were involved in those great successes (media interviews, service club appearances, etc.)." While he approved and arranged those appearances and meetings, he also ensured that the focus, the spotlight, was always solely on his people. He wanted the recognition to go exclusively to the members of his Air Station team who, through their superior performance, in the air and in the water, resulted in numerous lives saved.

An additional memory: the Coast Guard's 295-foot Barque EAGLE made a visit to Savannah during his tenure here, permitting invited guests to travel aboard this majestic sailing vessel, from Tybee Island (GA) several miles into the docking at Savannah's downtown riverfront. Commander Erickson made the trip in with his seagoing brethren, giving him another vivid Savannah experience. "Standing on the deck of the EAGLE, and watching three helicopters from my unit do multiple low passes, giving those on board ample photo opportunities, and just knowing that those Coast Guard pilots were having one of the best days of their life, that was really special," he recalled. Even the federal government shut down (January 2019), with its temporary lack of pay for personnel, made his list of Savannah memories. "That was hard, but we came together as a family. So, even though it was tough for our folks, there were some positive memorable moments during that period," he said.

And there was one public relations event while serving here in Savannah that remained firmly in Erickson's mind. It was the time he flew a local graduating senior to a reception at his school (May 2019). That student was Tyler Bland. His dad, Commander (retired) Chad Bland, a good friend of Erickson's, was a fellow Coast Guard pilot, a classmate of his at Purdue, and who had been in the same C-130 unit together at the Aviation Logistics Center. Tyler was graduating from Benedictine Military School ("BC") in Savannah and, most importantly, had been accepted to attend the Coast Guard Academy, the very reason Erickson was permitted to take him on the flight! He picked up Tyler and flew him from the Air Station to land right on the school grounds, the first time a helicopter had ever done that. As he landed, "the entire school was chanting his name. He got out of the helicopter and fellow students picked him up over their heads, and everybody's yelling 'Tyler Bland, Tyler Bland.' It was a joy to see them embrace Tyler as if he'd been there his entire life, even though he'd only attended for two years (CG dad had retired and moved to Savannah to work for Gulfstream)," remembered Erickson.

Then they all went into a near-by campus building for a brief ceremony, at which time Erickson presented Tyler with a certificate acknowledging his acceptance to the Academy, all the more special for this senior, since his school had a very definite military emphasis and tradition. Erickson then returned to the helicopter with Tyler and flew back to the Air Station. He later learned from some of the teachers that the event had been a very unique and exciting one, and that the students would be talking for some time about the day a Coast Guard helicopter actually landed at their school carrying a very special passenger, who was one of their own. Erickson hoped that maybe one day, one of those BC freshmen might recall that moment, and consider a rewarding career as an aviator in the Coast Guard.

*CDR Brian Erickson shakes hands with Mr. James Clark, President of South Carolina State University during a static display and signing of a MOU between SCSU and USCG to increase partnerships.*

Now, shifting from the present, and requesting that he look back to recall and summarize lessons-learned throughout his exceptional career-to-date, Commander Erickson was asked if he thought great leaders were born or made? "I think they're made," he responded. "Some people may not be capable of great leadership just because of the way they're wired. I think the right kind of influences in their up-bringing, and then along their pathway as adults, are probably what creates top-caliber leaders. Simply put, great leaders learn how to become great leaders from great people." Erickson went on to say that "great leaders routinely possess great character, which includes integrity and responsibility, all elements necessary to comprise a person's solid moral compass and positive outlook. And, as such, those are the leaders who are willing to take calculated risk. Failing to do that can often cause a lack of innovation, even success, within the organization and its missions."

With his defined set of great leader traits in mind, the Commander was then asked to think back to someone he had worked for, or with, who exemplified those qualities. Without any hesitation, the name Jason Tama came to mind. Captain Jason Tama is currently the Captain of the Port and the Commander of Sector New York, "the Super Bowl of Coast Guard Sectors," said Erickson. "More flag officers come from Sector New York than from any other sector assignment, a testament to the caliber of those selected to lead that demanding sector," he added. "Captain Tama is one of the smartest leaders I've ever met. He has an uncanny ability to recognize what's important and what's not, and he effectively separates the two. Additionally, he has a great sense of humor. He was the 'fun-est' boss I ever worked for," recalled Erickson. His job, back then, was demanding and required long hours, but he (Captain Tama) "made it a really good time."

Captain Tama was Erickson's supervisor during much of his time in the Office of Budgets & Programs at Coast Guard HQ in D.C. (2015-2017). When Erickson was asked for the key advice from Captain Tama that he most remembered, it was this: "Everything's gonna be OK." "Getting excited about something that doesn't matter, he said, is just noise," recalls Erickson. "The times I think about him are when my staff comes to me panicked over something. My reaction back to them is a very calm, non-panicked one. That's my way of attempting to calm them down, so that we can all rationally think through the problem and find a solution. Captain Tama was really good about doing that. Even with very important things back then, that to me, were critically important, like potential budget reductions, or dealings with Congress or late reports, he just made us all relax, calm down, and remind us that everything's gonna be OK. Let's get through this, he'd say, and figure out what needs to be done," remembered Erickson, reflecting on the obvious impact that important advice had on him, and especially so during his command.

Savannah was Erikson's first Coast Guard command. He was asked what he would miss the most. Without hesitation: "The joy of my peoples' successes. I honestly enjoy seeing my teammates reach their dreams, achieve their own individual definition of success, whatever that may be. "And, honestly, that's whether I've had any impact on it or not," said Erickson. "And I'll miss the comradery, the military comradery of an aviation unit," he added on a concluding personal note.

The next career step for Commander Erickson will be one-year of study (2020-2021) at MIT's highly regarded Sloan School of Business in Boston, to earn his Master of Business Administration (MBA) degree. You'll recall that he had already earned a master's degree at Purdue University several years earlier. This MIT tour is considered to be a senior service school assignment. Once again, as with Purdue, he had been competitively selected, becoming the only Coast Guard officer chosen in 2020 to participate in this prestigious MIT Sloan Fellows program.

As indicated, Commander Erickson has been selected for promotion to Captain, with the planned pinning-on of that higher rank in August (2020). Following his MBA year, he anticipates that his next assignment with the Coast Guard would most likely be a Headquarters (D.C.) staff position, but if there is a service need, an O-Six (Captain) command at a larger Air Station having multiple aircraft types. If the staff assignment comes first, then with his distinguished record, Captain-level command would most certainly follow.

During his command time in Savannah, he feels that he "tried to be really people focused, making sure that they were given all the tools and all the opportunities to succeed." And, that he did. Captain Erickson leaves command here with, he estimates, 35 lives saved or assisted, and a total of 2,600 Coast Guard flight hours (plus 2,500 civilian hours), during his service years to-date. He is a bright, highly respected officer, pilot, and leader. We wish Captain Brian Erickson all the very best as he continues with his impressive United States Coast Guard career.

**UPDATE**: Captain Erickson is currently the Deputy Director of the DHS Presidential Transition Office in Washington, DC. He plans to retire from the U.S. Coast Guard on July 1, 2025, after a little over 29 years of service to our nation (plus 4 years at the Coast Guard Academy).

*Captain Brian C. Erickson*
*United States Coast Guard.*

# CHAPTER 6

## LTC Bill Golden, U.S. Army, (Ret.)
## Iraq/Afghanistan Combat Pilot and Commander,
## 3/160th SOAR ("Night Stalkers")

William (Bill) Golden was born (1969) and raised in the small town of Dixon Mills, Alabama, situated "in the middle of nowhere," population "a few hundred," located about two-hours north of Mobile. Surrounded at home by a whole lot of timber and cattle, this one-time self-described 'farm boy' would grow up to fly and command some of the most sophisticated combat rotary-wing aircraft in the United States Army inventory.

Bill Golden attended Middle and High School in Sweet Water, Alabama. He played baseball and basketball, no mystery to the latter as he already stood 6' 3" making him a popular and effective stand-out. Small town, very normal small school (1-A) which, predictably, was the center of focus among the Sweet Water residents. However, the center of focus for Golden and his male friends was the "Ag Shop" and teacher, Mr. Phillips. "We all tried to spend as much time each day hanging out with him, and as little time as possible hanging out with our Algebra & English teachers (!)," he recalled.

An awareness of our military would, indeed, enter his life, but with no real impact during his high school years. His dad had, in fact, been drafted during the Vietnam era, but he never received orders to the combat zone. His uncle was also drafted into the Army, but he did end up deploying to Vietnam. Those family experiences were all he really knew about military service, and again, the notion of joining the Army never entered his mind in high school.

In fact, he not only had no intention of joining the military after graduation, he also had absolutely no desire to go on to college! But, regarding the latter, his dear and concerned mother had other ideas for young Bill. Amazingly, she took it upon herself to go to his high school, meet with his guidance counselor, then fill out a college application for him, submit it, and it actually ended up gaining not only his admission, but even a partial scholarship! The latter, perhaps not totally surprising, since "without studying much," he had achieved good grades in school. But prior to Mom Golden's unexpected initiative on his behalf, his announced intention was to go to work at the near-by paper mill along with his dad (who hoped that would happen, as well). "Yep, live on the farm and work at the mill, since that was the big industry down there," remembered Golden.

The college scholarship was to Livingston University, a small liberal arts school (since renamed the University of West Alabama). For the first year (1988), he commuted by car with a high school friend. After a year on the road, he decided that his commuting college experience was "just not rich enough," which he admits might have had something to do with breaking up with a girl! So, he moved to Livingston University and lived that next year with three guys in a trailer, for that "richer" college experience! Well, that enticing living arrangement about did him in as a collegian, so after a year of crowded trailer life, he knew, at last, that he really did have to get serious about college and transferred to the University of Alabama. He began his studies there in the Fall of 1990.

And now, finally, the idea of military service enters his mind and life, but almost by unintended proximity. At Alabama, he would share an apartment with a young man named Chad Ward, whom he knew from Livingston University. And here's the key that would now unlock Bill Golden's military door. You see, Chad Ward had moved on to the University of Alabama with the definite, and primary, intent to become an Army ROTC cadet. So committed was he, in fact, that the summer before beginning at Alabama, he had gone off to the National Guard's basic training program (SMP), providing him with some introductory Army experience and a bit of scholarship money.

So, Chad comes home from training, school begins, and "conversations about the military" inevitably begin, as well. It didn't click with Golden right away. He was a self-described (and likely over-stated!) "long-haired, trouble-making rebel," whose real goal at the time was to graduate and go on to law school. Before long, however, Chad, thankfully, convinced short-lived- "hippie," Bill Golden, to go with him to an ROTC event. "So, I go and meet a bunch of the guys and I really have a lot in common with them. Right away, I realized that I liked these guys. They were just good people," recalled Golden. With few new friends yet, since he'd just moved to the university, these young men quickly became a natural and comfortable source of friendships. Before long, he had joined with some of them on an ROTC-based competition team, named "Ranger Challenge."

He started training with them and, along the way, met all of the ROTC unit's cadre, the active-duty Army staff assigned to the unit. Along with that, perhaps a natural progression, since he was now spending so much time around both the cadets and the unit's supervising soldiers, he began taking some of the ROTC program courses. Before long, he was asked if we would be interested in attending the ROTC basic camp that next summer. That would consist of eight weeks at Fort Knox, Kentucky, and if he did really well in that program, he would be able to earn an ROTC scholarship. Golden did attend the summer program and ended up doing <u>very</u> well. How well? Oh, just becoming the very top student graduate in that particular camp cycle! His early upbringing had helped…a lot. Said he: "I'd grown up in the woods! I could navigate. I could patrol. And, physically, I had the ability. So, to me, it was a fairly easy fit."

Upon return to campus, he met with his cadre who then offered him a two-year, full-ride ROTC scholarship, if he was interested. He replied that he was definitely interested, but "I really want to go to flight school." By then, academically, he had become disillusioned with law, and had moved his undergraduate major over to corporate finance/investment management. He did decide to accept the scholarship, requiring him to sign a contract with the United States Army, with two-years of study left on what had now become a five-year university program, since he had lost hours when he transferred from Livingston.

That next summer, it was time for ROTC advanced camp, where he, again, did extremely well, becoming the top cadet from the Honor Platoon achieving the top score (a "5" on their grading scale). Only 10% of the 40 cadets he had trained with would achieve that top score. Now as a senior, based on his summer performance, he was promoted to serve as Executive Officer for Alabama's ROTC battalion. That put Golden in charge of the unit's Ranger Challenge team. And, under his leadership, that year, his team won the Ranger Challenge for the entire region, a goal long sought by his unit. Then, in December, he was informed that he had been 'picked up' for Army Aviation! Typically, only one cadet out of the entire senior class gets selected to go forward in aviation. In this case, for Golden, it was a combination of his top achievements and the encouragement/mentoring of the ROTC's professional military leadership.

Meanwhile, that senior year at Alabama, in the midst of his growing accomplishments, as those college years had become ever more serious to him, he had met the young lady who would be the love of his life, and so prior to leaving for his advanced camp summer training before his senior year at Alabama, he made time for one more vitally important life-time achievement: he married Carey! So, as he began his senior year, he had a beautiful wife and had now become the ROTC unit's battalion commander! The battalion won many awards that year, with Senior Cadet Golden in command. Prior to university graduation, Golden's first son, William, was born. Then this great, young, growing family packed up and headed to Army flight school at Fort Rucker, Alabama!

Bill Golden began flight school in November 1993. The Basic and Initial Entry Rotary Wing Courses took about a year. From there, after a bit of a wait, it was off to Advanced Flight Training. Golden was asked if he wanted to go forward flying fixed wing or rotary.

Without hesitation, his choice was rotary: "I wanted to fly Apaches. I definitely wanted to fly attack aircraft!" He finished #3 in his advanced class, and as a result, received approval to fly his first aircraft choice: the Apache!

*2LT Bill Golden flying*
*AH-64 Apache, Ft. Rucker 1995*

And so, his specific training on that airframe began. Pretty quickly, however, he began to have Apache doubts. "That's about the time I came to the conclusion that I don't know whether I'm cut out to fly Apaches. It was really difficult to fly this thing with one eye!" That "eye" reference was to a part of the flight program called the 'bag phase.' The first few weeks, he was taught to simply fly the aircraft, VFR, etc. 'And then, they put you in the 'bag' and it's a miserable, miserable experience. The windows are covered in foil, and the only way you can see is through a one-inch glass in front of your right eye. So, your whole world is made to exist just three feet below you and nine feet in front." And the reason for doing this? "Making you rely 100% on that little right-eye optic, and you have to learn not only to look through it, but also to read all the symbology. It's really, really difficult," remembered Golden.

*2LTs Bill Golden*
*and Bryce Dudley*
*Flight School Stick Buddies*
*Ft. Rucker, Al 1994*

But, as he recalled, it's kind of like learning to ride a bike. "You hate it, you hate it, you hate it, and then one day you get it. And true to what they say, you know, just stick with it and you'll get it." He stuck with it, and he 'got it'! Golden completed his training with an assignment to 18th Airborne Corps at Fort Bragg, North Carolina, to fly Apaches with the 229th Attack Helicopter Regiment there, "which was the best assignment I thought you could get. I was very proud to be going to Fort Bragg."

First-Lieutenant Golden arrived at Fort Bragg in June of 1995 and was given an Apache platoon (four aircraft) right out of the gate. He flew with his four-aircraft group for a year-and-a-half, and all is progressing well.

*1LT Bill Golden with his son, Blake*

At that point, he was 'put up' for pilot command (rear seat of the Apache). As he made that advancement, he and his group were notified that they would then be going to Bosnia! Amidst a leadership shuffle at that pre-deployment point, Golden was named a 3-5 support platoon leader, with the mission of delivering all the fuel and ammunition for the Apaches throughout the forward battle area.

So, he deployed from Fort Bragg with a platoon of 40 enlisted men and women, and perhaps as many as 20 trucks and trailers, with the job of moving the needed combat supplies from Taszar, Hungary down to Camp Comanche, Bosnia and Herzegovina. "And we do it, we do it well, we get it all sorted out, and we get it all allocated. We do all of that and we shoot a ton of helicopter gunneries. And I also get to fly as an aircraft pilot/commander," he recalled.

Golden was also quite proud that, unlike some of the other convoy leaders, he never got lost amidst those confusing, direction-less dirt roads, feeling like he had the best job in the world, getting to drive through the Bosnian countryside. And never getting lost was all thanks to Mother Golden, the very same wonderful woman who cleared (pushed!) his pathway to college. You see, Mom had purchased a Garmin moving-map GPS, which was cutting-edge technology at the time. She sent it over to him, so he'd know where he was at all times. And thankfully so, since the primitive road markings consisted of trees and an occasional rock with a direction arrow painted on it! He even flew with it in his Apache, since they didn't have the moving-map assist back then. Proudly, he saved it, and still has that Garmin device keepsake to this day.

Golden has some really good memories of his time on that deployment. "I learned a lot there. About accountability, responsibility, trusting your people, and the key value of having really good people. And, again, just basic trust." That assignment lasted six-months, during which time he was able to fly, and drive, all over Bosnia. And Bosnia would turn out to be the site of his promotion to Captain! He returned to Fort Bragg in Oct. 1997.

*CPT Bill Golden, Camp Comanche, Bosnia 1997*

Then, after being back at home station for a few months, out of the blue, or more accurately, out of the 'green,' Army Aviation Captain Bill Golden got orders to attend the Infantry Advanced Course (February 1998). What? Why would an established Army aviator be sent to an infantry school? "I didn't want to go," he recalled. "But I guess I was in really good shape, and they figured I had a really good profile record, so the Army decided I was going to go to Fort Benning, Georgia." Thankfully, his reaction upon completion was quite different from his first impression: "Turns out, it was the best thing I could ever have done! I mean I ran myself to death out there. And I realized that I was no longer the fastest guy in the company. Those infantry guys were exceptional (e.g., running/physically fit). But fortunately, I could hold my own!" Golden did so well, in fact, that he ended up being a class leader, which was a really unusual achievement for a non-infantry guy. "I actually read all the assigned books and I read the doctrine and, best of all, I understood it (good thing, because in the end, he had to know it!), while many of his classmates, he remembers, apparently chose to ignore it! So, apprehensive at first, Captain Golden ended up doing "really well" at the infantry school.

His next stop (his post-infantry school 'reward'?) would come in August 1998, when he received orders to Korea (unaccompanied!). Meaning, South Korea and without family! Awkward, in that, by now, he and Carey had three very young sons. However, as sometimes can happen with command connections, Golden's good friend and colleague from Fort Bragg, LTC Tony Crutchfield, was then the battalion commander where Golden would be stationed in Korea. So, with command approval, he was ultimately able to bring his family with him to Korea for that several month assignment! Fortunately, as well, in that all of the company commanders had their wives/families with them, so having Carey and the boys there wasn't an exception! Upon arrival in the Republic of Korea, he was given command of an Apache company, in the Second Infantry Division, with about 50 personnel and eight aircraft.

For Golden, this turned out to be another, as he put it, "phenomenal experience." A key reason was that he was then working with many of the captains who had trained with him at the Infantry Advanced Course! "They're the guys on the ground that I'm talking to. I know them all personally, so it was a very different experience because of that," he remembered. An additional highlight was that a couple of his officer counterparts (S-3's) were Task Force 160 aviators. Golden had been hearing about the 160th Special Operations Aviation Regiment (SOAR) for several years, and had decided that if the opportunity came along, he definitely wanted to move in that direction. With others around him having already assessed for the

Task Force, he decided that then was the right time. "So, I began to get myself ready mentally and physically (including doing a lot of swimming), along with considerable flying, including doing so with goggles, since nighttime is when the 160th flies its missions."

So, in the Summer of 1999, his preparation completed, Golden went to assess for Task Force 160. For him, even just approaching their post entrance made a real impression on him. "I definitely realized then, when heading through their gates, what a special unit I was walking into," he recalled. Golden assessed, did well, and was selected to join the Task Force. But, as it turned out, there was a bit of a problem. When he called his branch assignment officer back in Alexandria, Va to indicate that he had successfully assessed for the 160th. The Army had a different move in mind for Captain Golden.

Seems he had been forced to choose a next assignment before he had assessed for the 160th. And that assignment, completely different from his then-immediate goal, was to be an Observer Controller (O/C) at the Joint Readiness Training Center at Fort Polk, Louisiana for two years, after which the decision would be made about moving him to the 160th. It was the old 'be careful what you ask for,' but recall, that he had been compelled to make that next assignment choice prior to deciding on the 160th for his ultimate service direction.

Although initially unhappy about that assignment, it actually worked out quite well. Recalled Golden: "It was another one of those things that I just didn't want to do, but as it turned out, it was probably one of the best assignments I'd had. I really loved being there!" He was able to fly a lot while at Polk, accumulating about 450 flight hours in a small OH-58 Alpha Charlie scout aircraft, and almost 75% of those hours were with night vision goggles, perfect preparation for his planned future with the Special Operations 160th. "I learned so much flying that aircraft. And I got to go evaluate Apache companies, and OH-58D companies, as a company level evaluator," he recalled. After 16 months, he was able to at least move over to Special Operations aviation work, as the plans officer for the 160th, whenever they came down to Polk for pre-war training. In addition, while there, he got to attend "a ton of schools," one of which qualified him to control jets in the air for ground fighting integration.

So, then, one morning early, as was his normal routine, he was in Fort Polk's gym working out. The date was September 11, 2001. Then, unexpectedly, as it was for all of us, Golden saw something on the overhead TV screen that would forever be riveted in his mind. "I'm on the treadmill running, and I see the planes hit the building!" Struggling to get to the 160th at the time, he had been getting ready to go assess for a different Special Mission Unit. But once those twin towers in New York came down, and the Pentagon was hit as well, training pretty much stopped. "At that point, I got on the phone with my assignments guy and let him know clearly that I wanted to go with the 160th. I want to go!" said Golden, leaving no doubt about the seriousness, and the urgency, of his request. "My last bit of flying was about to come to a halt," he remembered. "If I don't get there now, I'm about to go fly a desk!"

Within the 160th, the established practice is that senior officers are permitted (expected) to do the same jobs that were done by lower ranks in regular Army aviation. Given the urgency, both from Golden, and for our nation at that critical juncture, at that point, things moved quickly for him. "Within 30-days, I have orders, and within 30-days later, I'm there!" In November 2001, Golden moved his family to the headquarters of the 160th at Fort Campbell, Kentucky, quickly bought a house, moved in, and was set to begin his long-desired and prepared-for 160th Special Operations Aviation Regiment (SOAR) career. First step, Green Platoon.

Green Platoon is all about learning to do the basics, at a higher level. "They teach you how to fly, shoot, swim, stress physical fitness, and you learn battlefield medical triage. And they teach you how to do it all exceptionally well," remembered student-again Bill Golden. For instance: "They don't teach you how to navigate with systems. You learned how to navigate with a clock, a compass, and a map." In terms of aircraft, he would now be flying MH-6 'Little Birds,' a McDonnell Douglas airframe, MD530F, specially modified years ago by the 160th to fly specialized missions. "These are urban fighters," said Golden. "Four of them in a real tight row, go down a street, with guys sitting on the outside (remembered from 'Black Hawk Down'). That's the company I'm assigned to and trained with (1/160th/Task

Force 160/packaged per mission & attached to a ground force) at this point," he recalled. It was then April 2002.

*MH-6 Little Bird*
*Baghdad, Iraq*

At that time, the War in Afghanistan was underway, but there was no role then for 'Little Birds.' From the standpoint of the 160th, "it was a Chinook and Black Hawk war," remembered Golden. Then, in August 2002, Golden began his first war-time deployment, this one to Saudi Arabia, as a liaison officer, operating in a Combined Air Operations Center ("CAOC"), which is the control point for all of the Air Force, Navy, and Marine aviation assets. During his ninety-day assignment there, he was "working all sorts of coordination missions on the ground. Not flying at all. Twelve hours on/twelve hours off. I learned a ton. I learn how we run air battles and who controls them."

Later that fall, mission planning began for the invasion of Iraq. "I get put on the package that's going to deploy to Iraq," said Golden. "As I recall, there were a lot of unhappy people over who got picked and who didn't. I would be commanding the 'Little Bird' assault package that went to Iraq." That aircraft simply hadn't been the right fit for Afghanistan. "But Iraq didn't have the same type of terrain to deal with, it wasn't that high above sea level, and there was a lot of urban fighting. And so, we were fighting in urban areas, towns/villages, and the 'Little Bird' was the perfect platform to do that mission," recalled Golden.

"My first mission to Iraq was in support of a special operations ground squadron that was having to do a deep penetration into Iraq, because Turkey would not allow entry for our forces," recalled Golden. "They launched a special-ops squadron early, into the western part of Iraq, to try to prevent Saddam Hussein and his family from escaping to Jordan and prevent the forces in Northern Iraq from reinforcing south. They went up into Tikrit and severed the main road so that we could isolate Baghdad, and so that the Marines and Third Infantry Division could make it to the capital city."

*MH-6*
*Iraq*
*2005*

During this phase, he and his unit were out west in the desert handling missions at night, while sleeping, and camouflaging their aircraft to prevent detection, during the day. Then they changed out with other units, loaded up, and headed to Baghdad! By this time, our Marines and the Third Infantry Division soldiers had made it there to Iraq's capital city. The Americans had cratered the main runways at Baghdad International Airport, to prevent jets from landing and taking off from there, so when they flew into Baghdad on MC-130 Aircraft, they had to land on the Parallel Taxiway. "And then we immediately unload began looking for Americans who had been taken hostage by Hussein's forces."

North Eastern, Saudi Arabia had been the forward stationing base for Golden's unit. Now, he moved his guys on into Baghdad International as the fighting intensified. "We arrived and went into this blown-to-pieces hanger and that's where we're living, cots or couches on the floor. Ground forces and aviators, we're all in there together," said Golden. "It was an interesting time! We'd wake up and head out on a mission to try to find those hostages (assisted by U.S. intelligence). While assigned to the 160th, It was the first time I'd ever worked with a mechanized armor squadron (Army 3rd ID)." And he quickly discovered that 160th aviators and Abrams tanks &

Bradley fighting vehicles function on different timelines. Very different timelines!

Golden's aircraft were out on station way before they could hear and see the approaching armor unit. "We arrive in our Helos at the time the mission brief specified we should arrive, with snipers onboard my platform, flying around the suspected location looking for any movement or sign of people. Just flying around wondering where our tanks are when, out my right door, a giant explosion goes off. Well, it's the Cav. shooting rounds into an already quite destroyed Iraqi (old Soviet) tank! And they kept on firing, while I concentrated on getting the heck out of the way! That was one of my first 'welcome to working with armor' experiences.

They turned out to be great professionals, but those first few missions were really a big learning experience for us," he remembered.

And that started repetitive rotations to Iraq for Golden. "I'd go home for a month or two, then go back for a month or two. Until, ultimately, I came back to the U.S. to attend Command & General Staff College (Summer 2004) for a year at Maxwell Air Force Base in Montgomery, AL.". (He was promoted to Major in June of 2004 at Fort Campbell, KY.) His mother (in Alabama) passed away there during that period, so it was fortuitous for him to be back in Alabama to help his dad during that sad and trying family time.

One year of schooling passes, he graduated, and then it was back to Fort Campbell for Major Golden and his family. There, he took command of the same company (vs. Captains in the regular Army; in the 160th, Majors command companies) where he had previously served as a platoon leader (A Company/1-160th). Two years followed, with the same shuttling back and forth to Iraq. "And we were doing all kinds of innovative things, finding the enemy wherever they roamed and hunting them in ways not before utilized" said Golden. "We developed all kinds of dynamic tactics for taking down targets. What happened was that the enemy got savvy to the fact that we had given them the ability to move around in the daytime and communicate freely. ('Little Birds' had generally flown only at night)!

So, we changed our tactics. We began to dynamically target them while they were moving," he recalled. "Primarily going after high-value targets. The type of individuals was bringing in high-casualty-causing IED's that had a core capable of penetrating our armored vehicles, including tanks. So, those high-value individuals became part of our focus. And we got really good at it! It was a tough blow to them, because it resulted in very dynamic engagements. We interrupted and tracked the way they maneuvered. They (enemy) were forced to get very quiet. They stopped being able to talk to one another. Which made it difficult for them to control a cell. It just put fear in their hearts to drive, to move, to talk, to do anything, and we just significantly disrupted them," he remembered.

Golden spent two years in company command. Shuttling back and forth between home, Iraq, and Afghanistan, "over and over. And we spent those years just honing our skills, and getting very good at our jobs," said Golden. In 2007, he came out of company command to become the Executive Officer (XO) for the First Battalion of the 160th at Fort Campbell (about 800 soldiers), during which time he also became the Task Force Commander overseas multiple times. "And that means the decision-maker for missions along with the ground force. You are the final word. You may have to say "No" on some missions, and that's really tough with ground forces. So, you've got to have the credibility to be able to do that. And you have to be right!" recalled Golden. "Those were some tough years, we lost our teammates, but there was a lot of learning, and we got really good at what we did," said Golden.

In 2008, while still with Special Operations, he was given a two-year assignment, state-side, to Alexandria, VA, with the Special Management Division. He became the "assignments guy," said Golden, meaning that he had to designate prospective 160th soldiers for the Green Platoon, working directly with Human Resources Command, which also included solving SOAR recruiting issues during wartime. "The entire Army was deployed. Everybody was hurting for people. And there weren't enough Army aviation brigades to adequately source prospects for the 160th. We generally don't take guys out of flight school, so candidates would have to come from within the conventional Army. But often those commanders wouldn't permit their aviators to assess for the 160th because they were too valuable to them right where they were," said Golden, thinking back.

"So, it became a bit of an adversarial relationship. We were all fighting for the same resources. They had their missions to accomplish, and we had ours. Fortunately, the Army G-1 determined our recruiting needs to be the top priority," recalled Golden. Although this was an assignment he never wanted nor sought, he did end up learning things he never knew. So, for him, it actually turned out to be "a great two years." But initially, he had made it quite clear to his Commander that this "was definitely the last job he wanted." And without missing a beat, his Commander gave him this classic response: "Good, then you're the perfect person for it!" Among the positives of this two-year assignment, while working in Alexandria, he was promoted to Lieutenant-Colonel (Spring, 2010).

Following this important manpower job, LTC Golden returned to Fort Campbell, knowing that he had successfully passed the boards and had been selected for up-coming command of the 3rd battalion of the 160th. There would, however, be a one-year wait. At Campbell, he began transitioning into Chinooks, following all of his experience flying Apaches and 'Little Birds.' So, he spent the next year going through the MH-47G Chinook qualification course, completing his pre-command courses, and serving as the Regiment's planning officer for fielding their new model Black Hawks (MH-60M).

In June 2011, he arrived at Hunter Army Airfield in Savannah, Georgia as the new battalion commander of the 3/160th SOAR ("Night Stalkers"). And while our nation's military continued to shuttle deployments between Iraq and Afghanistan, the war focus at that time, was clearly shifting more totally toward Afghanistan. As a matter of fact, the last three Black Hawks left in Iraq that year were actually from Golden's 3/160th battalion. And with America's more dominant emphasis now turning to Afghanistan, along with assuming his battalion command, just one month after arriving at Hunter, LTC Golden was then deployed to that country as Task Force Commander (July 2011).

*LTC Bill Golden in Kandahar Afghanistan
with the ARSOA Task Force, 2011*

"So, I go over there with the team, and went directly to Kandahar and spent about a week flying with my guys. That's what I wanted to do. I did missions with them, now flying Chinooks," said Golden. And he vividly remembers his very first Chinook mission flight there. He was flying out of their operations site near Shank, Afghanistan, a Forward Operating Base (FOB) that sits at about 7,000-feet above sea level. It's way up in the mountains with very rugged terrain, located over in the east-central part of Afghanistan, with, reportedly, some really dangerous bad-guy areas in that region, in fact, well known to be "rife with Taliban and al Qaeda." Golden recalled that: "The terrain was unlike anything I'd ever experienced in my life. It's just mountains everywhere, with deep valleys and gorges running through them. And these are the very areas where U.S. troops are known to get shot at quite a bit!

"I get into the Chinook, and we taxi out and take off. And the guy (co-pilot) who's flying with me is probably overly trusting in my capabilities with this aircraft! Mainly, at this point, because I'd never seen, nor flown, in dark like this," he recalled. "Even just trying to keep up with my lead aircraft was difficult. I was concentrating on staying close to him, while not hitting mountains! And while listening to all the radios. There's just a lot going on, and I'm trying to freakin' keep up with the guy flying in front. All I can see are his IR lights, which are meant to be really subdued, so that people on the ground can't see them! And all the while I'm thinking, this is going to be one intense challenge. Meanwhile, the crew on board is going through their checks, getting ready to commence weapons fire, if needed. The guys are amazing, super calm, no one seemed concerned. And in my head, amidst the almost total blackout outside, I'm having this conversation with myself: Am I really ready to do this? Matters not, as we're soon nearing the release point, one-minute from touch-down. And, again, I haven't landed a Chinook in combat – ever – and worse yet, never in blinding dust. I've trained, but never done this for real! I can see our lead aircraft heading down in front of me. Then, suddenly, the dust kicks up and the lead Chinook disappears. At about 80-feet down, I can't see the ground anymore. I come all the way down to the deck, with the crew chiefs calling out what they can see. And I'm looking at the symbology, and I land that Chinook! Bumpy, but landed and stopped," remembered Golden, in amazing detail, as if it had just been the day before.

Upon landing, ramp down, the guys (special-ops soldiers in the back) run off to begin their ground mission. The Crew Chiefs then quickly raised the rear ramp, and he took off. He and his crew returned to their home field without incident. "We survived!" said Golden. "I had completed my first 'in-fill' Chinook mission in Afghanistan. I was feeling completely overwhelmed because of the total darkness, and the challenge of the surrounding terrain, but what I realized is that the training that we all get is far more than adequate to see us through. You know, trust your equipment, trust your people, trust your training, and it's going to work out fine," said Golden. "And that mission was just the first of many. I was really blessed to be able to fly a lot while I was over there." He would rotate back and forth, to home and war (deploy two months, then home six.)

During October 2012, came one of the more memorable missions that he personally flew as commander in Afghanistan, and one of several that he remembered very vividly. It was one that would take his team out near Bastion, Afghanistan. "I always felt like I wasn't the one taking the biggest risk as the aviation component. If they were willing to go in on the ground, we'd do everything possible to fly them in there. And this night, we were going to in-fill 200 troops, both ours and the attached Afghans, within five aircraft," said

Golden. This particular place, however, was one where our teams were always taking significant ground fire. In fact, the prior week, U.S. Marine aircraft had tried to fly in, but had been shot at so badly that they had to abort the mission. So, the Marines special operations folks called Golden to see if he and his team would be willing to do that in-fill mission.

*LTC Bill Golden (far left) and CSM Estevan Sotorosado (center) with MH-47 Crew prior to a mission in Afghanistan*

As he most often did when requested to assist in getting specialized troops into key operations locations, Golden told the Marines they'd do it. "Let us get the illumination and everything else right, and, yes, we'll take you in. So, we flew out to pick up the Marine Special Operators and Afghan troops. I had five Chinooks, one from every SOAR battalion. That's a whole lot of thunder! We take off with our share of those 200 Marines & Afghans in back, and on that flight in, we had Marine AH-1's with us, plus AC-130 gunships overhead, and ISR (Intelligence/Surveillance/Reconnaissance). One of our overhead platforms was listening-in on enemy frequencies. And they reported enemy ground chatter on the radio about "something going on." No question, they knew we were coming." This particular area, the Helmand River Valley, was home to ample river water, a perfect lush environment for growing the region's dominant cash crop: poppies! Because of the huge economic value involved, the Taliban and al Qaeda forces were militant about protecting the area from disturbance or destruction by American forces, whether from the air or on the ground.

As Golden remembered the mission: "It was about two in the morning, scheduled purposely, because if there's someone(s) down on the ground still awake at that hour, they're probably a bad human being! And so, as we fly down into this valley, we all split up and we're landing probably about a quarter of a mile from each other. Coming down, we're 'browned-out,' flying in through the giant dust cloud created by our landing approach, which kinda protects us, because not only can we not see, they (the enemy) can't see us either! Ramp down and the Marines run out of the back. Then we get set to take off. And as soon as I do, we spot a guy on the ground smoking a cigarette. It's around two in the morning. Gotta be a bad dude! That's when we start receiving ground fire. The first guy to begin shooting is my right door gunner and he's just hammering away on the minigun. About that time, I start flying out of the dust cloud, and then my aft machine gunner sees what my door gunner is shooting at, and he starts engaging, too."

"With the billowing dust, and the intense darkness, thankfully, the bad guys can't clearly see us. They're just aiming at our engine sound. Three of our four guns on board are now firing cover, as we come out of the dust, and I'm now flying low down that valley as fast as I can. My guys (gunners) are just smoking it, and we're coming out of there. Then, the smell of cordite hits me in the nose (flow of air in a Chinook goes from rear to front!). My eyes are burning. All five of our aircraft are now up and flying, with 'red ropes' (visual defensive gunfire) coming out of them. I think to myself, this is gonna be interesting. We're all trying to get out of this valley together (5 Chinooks), and we're all still being shot at from the ground, with machine guns and RPG's. Miraculously, not a single bullet, or RPG, hit our aircraft. We shot thousands of rounds. It's a tribute to their (gunners) ability to immediately pick up a threat, and then violently suppress it with overwhelming firepower. It's that firepower that keeps bad things from happening. I attribute our good fortune, again, to great training, great communication, great situational awareness, and those guys just being damn good gunners!" recalled Golden.

Following a combat zone troop "in/ex-fill," very often the 160th Chinooks would fly back to their staging area to refuel and "exhale"! And then stand-by for another flight assignment (very often at least two per night). But, when called, there were times when "we would literally fly pretty much all night," he remembered. That continuous tasking rigor, and taking on the toughest missions when requested, was fully reflective of the Regiment's long-standing motto and fervent belief: "Night Stalkers Don't Quit!" "I did a lot of other missions, but those two described are just a couple of the more memorable ones in Afghanistan. And after each, without fail" concluded Golden, "you're always mighty thankful that God was with us."

Reflecting back, then, on his time in command of the 3rd of the 160th in Savannah at Hunter Army Airfield (2011-2013), Bill Golden recalled that one of the things he loved about Third Battalion was that "you were king of your own domain. No one was down here from the 160th HQ (Fort Campbell), it was just us. There was complete autonomy of command. I just absolutely loved it!"

*LTC Bill Golden*
*CW5 Tim Denton (left)*
*CSM Todd Hedrick (right)*
*3/160th Command Team*
*2013*

While the Middle East warfare continued, "we also had a very focused training mission in South America (unit flew aircraft all the way there, a first). Fortunately, in the midst of it all, we had plenty of resources to train. We, of course, had guys continually deployed Afghanistan. We had great helicopters (Black Hawks & Chinooks). And we were blessed with highly trained and motivated individuals. And, recall, in my role prior to command, I had actually worked to resource Third Battalion, shaping the equipment & the people (Officers & Warrants), even before learning that I'd be coming here to command!" recalled Golden, thinking back. He also remembered the opportunity to consolidate his medics (18-all on flight status), and other staff units, in newer facilities at Hunter, making the coordination of activities and missions much more efficient.

Was there anything else that stands out in your mind about your Savannah tour? "Well, yes, definitely. The next thing I remember about being here was the people of the area. I had never been to a place in the military where area folks actually partnered with you, coming up and saying, hey, how can we help you be successful. I had never experienced that at all. So, it was the people we led, and the people who were around us in the community, that made that command experience so special and rewarding," he said.

"My goal was to leave the battalion as healthy as I could, as highly trained as possible, and with the best leaders we could possibly provide. Previously all of the battalion commanders had historically come from 3rd Battalion. After my departure, some key 3/160th officers went on to command other SOAR battalions. That's what I had hoped for, that was my goal, to develop super focused, driven leaders who could one day lead the Regiment. And that subsequent success was a source of personal pride from the leadership work I had put in," said Golden.

*LTC Bill Golden with his wife*
*Carey Golden*

Bill Golden is simply an outstanding example of impressive service, both in the military and well beyond in civilian life, focusing, as he has, beyond his demanding Bank of America executive work, on giving qualified veterans an important employment leg-up in transition and helping Night Stalker families thrive in a very rewarding and challenging assignment.

**UPDATE:** Bill Golden retired from Command of the 160th SOAR Battalion, and from the United States Army, in June 2013 after 20 ½ years of service.  He was selected for an internship with Bank of America in New York City and was immediately hired full-time upon completion. His continuing Bank of America tenure began as VP Global Equities Risk, rising rapidly to his current position as Managing Director of Global Market Operations.  He currently responsible for about 400 employees globally.

Golden also leads two internship hiring programs for veterans, with special focus on those who are interested in working full-time for Bank of America. Additionally, he was the Commander of the American Legion Post 754 (New York Athletic Club), focused again on helping veterans interested in financial services careers on Wall Street to secure entry internships. Along with all of that important involvement with job placement for veterans in New York City, Bill Golden is a Founding Director, now Chairman, of the Night Stalker Foundation, with a main focus on helping care for families by funding a program on being resilient and thriving, funding college scholarships for Night Stalkers, along with support efforts for Night Stalker Gold Star families.

# CHAPTER 7

## World War II B-24 Combat Pilot
## Lt. Paul C. Grassey
## U.S. Army Air Forces

Following high school graduation, and awhile later at age 19, Paul Grassey had important decisions to make. Although attending college was part of the mix, the reality of the times really came down to either enlisting in our military or waiting to be drafted! He spent time talking about options with many of the same buddies he'd spent years playing sports in high school. He remembers asking the father of one of his good friends what he thought he should do. Turns out this gentleman had actually flown as an American aviator with the legendary French Lafayette Escadrille (1916-1918, prior to America's entry) during World War I. One weekend later, Paul remembers his friend's dad coming downstairs wearing his full WW I uniform with Captain's bars and wings. At which point, he announced to Paul and his amazed, but still undecided whether-or-not-to-join-up, friends: "I'm leaving Tuesday. You guys make up your own minds!"

This patriot WW I flyer had actually decided to serve his country, yet again, re-entering the Army as a pilot, and he would fly 9 missions overseas. Best of all, given his remarkable effort, he survived the war. That re-entry action so impressed Paul that it became his decision-point. Enough wondering. He decided, then, to enlist in the Army Air Force. Said he: "I made that decision because, to me, it was important. The world was on fire. I wanted to do what I could to serve and help my country."

*Paul Grassey in pilot training.*

Paul knew he wanted to fly for America in the war. Most of all, he wanted to fly as a pilot. At an earlier time in the war, the requirement for pilot training was a minimum of two years of college. As the war progressed, and U.S. air casualties mounted, the rules changed, the requirements lessened, meaning that he could now be eligible for flight training with just three-months campus time. And that he did, attending Lafayette College, taking some coursework but, importantly, also gaining 10 hours of flight time in a Piper Cub. Requirements met, and with additional initial training completed, it was off to Nashville (TN) for, among other things, placement testing to determine whether young Mr. Grassey would qualify for pilot training or, instead, join the war effort in the air as a navigator or bombardier officer. Paul scored well on the aptitude testing and was officially selected for pilot training. With those sequences successfully completed, in the 1943-44 timeframe, he graduated from flight school, pinned on his wings, and became an Officer in the U.S. Army Air Force.

*Paul Grassey (second from left, kneeling) with his B-24 crew before heading to England.*

From the air base in Charleston, S.C., then on to Mitchell Field on Long Island (NY) to pick up their newer model, the B-24(M). And from there it was off for Bungay, England (with refueling stops along the way) to join the Eighth Air Force's 446th Bomb Group and the air war against Nazi Germany, then already long underway. It was

December 1944, when Paul and his crew arrived to do their part against a still determined, powerful, and stubborn enemy. Paul Grassey would pilot a total of 13 missions in air combat through the still-hostile, war-torn skies of Europe. Throughout all those missions, miraculously, he doesn't recall his aircraft ever being hit, by either ground fire or from the guns of a Germany fighter. Clearly, Paul and his crewmates had an impressive Guardian Angel!

Ironically, the closest he and his crew came to having their aircraft fall out of the sky was over the North Atlantic on their initial flight from America over to England! Despite refueling at the prescribed locations along the overseas route, for some reason, at some point, Paul became aware that they were running low on fuel. They were then seven hours from re-fueling in Greenland, but the instruments were telling Paul they had only six hours of fuel left! Needless to say, that math wasn't working.

Now, ordinarily, the pending problem could be solved fairly easily. The much-improved M-model aircraft they were flying was equipped for the journey with a so-called "Tokyo" tank carrying extra fuel for just such a potential emergency. All that had to be done was for the crew's Flight Engineer to transfer the needed fuel to the main engine tanks. Basic procedure, one would think, except Lt. Grassey quickly learned there was a bit of a problem, when his Flight Engineer confided to him that he couldn't make the transfer. When asked why, he stunned his pilot by responding: "Because I don't know how to do that!" Likely just about the last thing a pilot on the verge of a crisis would ever want to hear. What to do, and do quickly, in this well-before-Google era (!). Well, they got on the aircraft's radio, indicated their dilemma, and out there, somewhere, another Guardian Angel, this one in uniform, answered. Seems there was a booster pump located behind the cabin door, with a switch on it that had to be set to the 'on' position. That life-saving instruction in hand, the fuel transfer was successfully and thankfully made.

*Lt. Paul Grassey*
*WW II combat pilot.*

Worth noting that, at that time, out over the ocean, they were cruising at about 13,000-feet, flying through December winter weather, and as mentioned earlier, they had been several hours away from their next re-fueling stop. Serious stuff. Paul remembers that four other B-24 crews were lost, and never found, making that same North Atlantic transit. Hard to think that the lives of his crew of ten young Americans could very well have been lost to the unforgiving seas of the North Atlantic without ever getting to their base in England, or flying a mission in actual combat, which they had all been trained for, and were anxiously expecting to do.

The other flight that, after some 75 years, remains riveted in Paul Grassey's mind turned out to be the mission that never was. Had it happened, it would've given Paul 14 total missions flown in combat. As it was, it was still very much flying in a combat environment. Following the morning crew briefings, Paul and his crew rode out to their plane, did their internal checks, and prepared for their turn to take off. B-24's, he said, required more ground speed for lift off than the B-17's, but, regardless, the runways all seemed to be the same length! Concern was always about clearing the trees which, with American bases in England, inevitably seemed to be always standing there as the runway pavement was coming to an end.

When they left England that day, the weather was clear and fine. Those conditions didn't last. Once over the continent, and with the remaining two-hour flight to Paris, the conditions had deteriorated severely, with heavy rain and lightning covering the flight path for hundreds of miles forward. As Paul recalls, there were now about 150 American bombers, 17's and 24's, circling later over the French capital, trying to both link up with their mission leads, while at the same time trying hard not to collide with any of that large number of aircraft circling and waiting around them in blinding weather.

At some point, the radios came alive with word from headquarters that, due to the weather, their mission(s) for that day had been cancelled. Disappointing for the aviators, hoping they'd have been one mission closer to completing their total and heading back home to America. But yet, a relief not to have to plunge forward through such dangerous flying weather, over territory still very much in enemy hands. The visibility was reduced to the point where even finding, let alone hitting and destroying, their assigned target(s), was in serious doubt.

And then their return flight to England got even more challenging, yet, since their plane's radar had, for some reason, gone out! Adding to the headaches, the rainwater was by then hitting the aircraft so hard that it was actually leaking through the pilots' front windshield! So, in the skies over Paris, in the midst of weather chaos, the circling bombers carefully made a sweeping half-circle back around to head, much sooner than anticipated, back "home" to eastern England. Now, absent any allied fighter support, which had also been called back, the bomb group would have to fly over Belgium, and its dreaded port city of Antwerp. Dreaded because it was still very much occupied and featured a sizeable number of German anti-aircraft guns ("88's") protecting this key German-held port. And to make matters even worse, particularly in poor-visibility

weather, along with those ever-lethal flak guns, Antwerp was surrounded by a large array of barrage balloons, whose steel cables could bring down a low flying bomber with one strike of a wing. This added threat, while Paul and the others were already flying lower than normal, in order to try to avoid alerting German fighters on the ground, whose radar was quite likely still working!

There were 23 B-24's flying in Paul's group. He wasn't certain, but he remembered that either three or four of them did end up being shot down by German ground fire. Fortunately, for

*Paul Grassey in flight.*

Paul and his crewmates flying this, as it turned out, <u>non</u>-mission (although very much still in combat), their aircraft did make it safely back to their home field in England. With noted relief and classic understatement, from this and all of his missions facing enemy air and ground fire, as Paul recalled: "Combat is 95% luck. There are just so many ways to get killed!" Fortunately for Paul, his luck in the air never ran out. Thankfully, it won out.

When it came to bomber piloting skills in combat, Paul Grassey was the proudest of his ability to maintain his aircraft in a tight group formation in the air. Tight formation meant flying virtually wing tip to wing tip, with only a few feet separating one B-24 from the one on either side. This was done to provide the most self-protection possible for our bombers from the attacking German fighters, by maximizing the concentration of our aircrafts' defensive machine gun fire. Tougher for an enemy fighter to pick off one of our bombers flying amidst that barrage of returning fire from our array of gunners (front/back/sides/beneath). Paul indicated that one time (at least), unhappy with the loose formations he saw among his bombers, pilot and leader, General Lew Lyle, actually flew right through the formation to prove his point that his guys weren't flying nearly close enough for maximum safety! With that dangerous and dynamic example, pretty sure General Lyle made his point!

Asked what sensory memories he recalled the most while flying within America's bomber armada through those death-dealing enemy skies, Paul remembered, perhaps most, the "awfully noisy" (but yet comforting!) sound of those four powerful Pratt & Whitney engines. Next, was the constant aroma (make that, smell!) of oil coming from those same sturdy and reliable big engines. Then, once in actual combat, along with keeping his head on a swivel to maintain safe distances in formation, while ever-scanning the sky for enemy fighters, he remembers the cascading sound of machine gun shell casings pinging off the steel-plated flight deck, just behind the pilot seats, as the flight engineer manned the defensive forward machine gun over-head.

In addition to having a longer range than their sister B-17's, the normal cruising speed for a B-24 was also faster. Generally, an advantage, but it did mean that they could not do formation flying with B-17's. Said Grassey: "If we were to throttle back in order to stay with them, we could end up actually staling out!" He did feel that the B-17's were more "air worthy, however, because they had that "big wing."

Not surprising for any sane person, when asked what scared him the most flying in combat, facing both ground an enemy aircraft fire, Grassey replied without hesitation: "Being shot at!!" And what quickly became quite apparent to him in that situation, he recalled: "They're actually trying to kill us!" He remembered feeling the most uneasy just as the action from below and above began. "I didn't worry as much while flying tight formation and the guns were firing. There wasn't time for worry then. Way too much going on. Were we scared? Of course. But upfront, at least we could see what was going on. The guys in the back of the plane didn't really know what all was happening. Maybe somehow that was better?" Speaking of the guys in the back, there was one gun position, he was glad wasn't his. "The guy in the ball turret. I wouldn't want that job for anything!"

Grassey did recall clearly the impact of those flack shells coming up and bursting around them: "The flack was brutal. Maintaining formation for protection, with that flack exploding all around us, all we could do was sit there, take it, and hopefully keep flying." With their bomb load finally dropped on target, the immediate requirement was to "re-trim" the now-lighter aircraft in order to regain that tight formation among the other 24's. Formation "was the best self-protection we had," remembered Grassey.

In flight, the two pilots would take the controls twenty-minutes on and twenty-minutes off, so that they could each stay alert on those sometimes very long flights, especially as American and British ground forces moved enemy troops and their defenses further and further back into Germany. In that regard, he distinctly recalled one particular mission deep into Germany. This time it was the ultimate Allied objective: Berlin. Grassey mentioned that there were somewhere close to 24,000 defensive anti-aircraft guns ringing that capital city. And here comes another classic Paul Grassey understatement: "The more guns, the more trouble you could get into!" He singled out that Berlin mission for its obvious heavily defended challenges: "That was a very tough mission. It was just total action." Sometimes they faced more "action" from the flack guns, while other times it was more from the enemy fighters. While they did have at least some basic pre-mission intelligence presented to them during their preparatory briefings, the reality of what they would face could never be fully known back then just from staring at a very large area wall map! Despite the best possible intelligence and pre-planning, as combat veterans, through the decades know all too well, the harsh reality of combat never fully reveals itself until that first actual contact with the enemy, on the ground, at sea, or in the air.

Thankfully, for Paul and Allied forces throughout Europe, the four-year war there for America's military (six-years for Britain, France & others) finally ended with Germany's unconditional surrender on May 7, 1945. From that point, Lt. Grassey and thousands of other American pilots assigned to the European Theater received orders to return to the United States and prepare to join their colleagues in the continuing fight against the remaining Japanese forces still battling in the Pacific. Thanks, ultimately, to the American-developed atomic bomb, and the massive Army Air Force B-29's capable of transporting and delivering them it on-target, the war with Japan ended, and it did so, without the necessity of Allied forces having to invade the Japanese homeland, saving the lives of countless Americans, other Allied troops, and additional Japanese. Thankfully, then, it was no longer necessary for Paul Grassey to join that fight. He was back on American soil and honorably discharged from the United States Army in November 1945.

Once back home, he enrolled at Lafayette College (Easton, PA) where, in 1948, he completed his war-interrupted B.A. degree. He then went to work for the Burroughs Corporation, a dominant producer and distributor of office equipment, then very much in demand, as the nation's businesses and industries experienced accelerated peacetime expansion, innovation, and productivity. Paul spent a total of 37 years with Burroughs, moving often, each with increasing responsibilities, finishing his career with the company in New York City as District Sales Manager, by then with about 500 employees under his leadership. He and his wife Nancy (a former Air Force nurse) then moved to Savannah, Georgia as their retirement home in 1988, for which this community has been continually grateful.

*Paul Grassey and Scott Loehr (Museum President & CEO) at lunch at the National Museum of the Mighty Eighth Air Force.*

Paul Grassey became associated with The National Museum of the Mighty Eighth Air Force (the command with whom he flew), as both a tour leader and Board member, during the entire 25-years, to-date, of this outstanding historic Museum's existence (1996-present). In 2013, Paul authored the book, <u>It's Character That Counts</u>, in which he described memories from his youth, wartime, and post-war business career, including key friends and mentors who were a positive influence on him, due to favorable, and memorable, aspects of their personalities and positive character.

In January 2020, Paul Grassey received the French Legion of Honor, a medal which since 2004 has been awarded to "American veterans who risked their lives during World War II during one of the four campaigns for the liberation of France." Coming to the Museum from his headquarters in Atlanta, Georgia, Consul General of France, Mr. Vincent Hommeril, made this special presentation in front of a large number of Paul's family members, friends, and a contingent of service members from the near-by 165th Airlift Wing of the Georgia Air National Guard. All were there to congratulate Paul for this high honor, reflecting a portion of his courageous actions and efforts during World War II.

*Paul Grassey at the Museum.*

In addition to his activities with the Museum, still very active and on-the-go, Paul Grassey is a much-requested public speaker, making frequent presentations to area schools, conferences, and civic groups. Along with speaking, he thoroughly enjoys singing patriotic songs and tunes from the 1940's, which with his very pleasant baritone voice, the audiences always enjoy hearing. Paul Grassey and his wife Nancy remain long-time honored members of the Savannah community. We laud, honor, and deeply respect the service that both have given to our great nation.

**UPDATE:** Interview sessions for this career narrative were conducted during November 2020. At age 97, following a brief illness, Paul Grassey passed away on Sunday, April 11, 2021. Rest well forever, dear friend to so many … a truly devoted, inspirational, and courageous Great American.

# CHAPTER 8

## "I Thought For Sure I Was Gonna Die!"
## United States Army Major General
## John D. Kline

That harrowing sentiment from one of the more intense combat mission experiences recalled by United States Army pilot, then-Major (now Brigadier-General) John D. Kline, as he fought to safely land his Black Hawk helicopter on a mission in Iraq back in 2005. No time or place for a rookie pilot, fortunately Major Kline was far from it, a highly skilled veteran aviator and unit leader. And, also fortunately, characteristic for that time in the Iraq War, no one was shooting at him! Ground fire, there, hadn't yet become a real and persistent issue. So why did he think this flight might be his last?

Because he now found himself flying in a desert condition, known as a brown-out, blowing, thick dust everywhere, and with it, darkness, effectively blinding his approach. "We were coming in and actually hovered for a while in the soup, as that major dust storm was really building up." While underway, most combat helicopter accidents, and related deaths, said Kline, were the result of either pilot error, treacherous mountain terrain (Afghanistan), and/or brown outs, just like the one he was then battling.

Beyond concern for his own safety and that of his crew, he was transporting eleven Army Infantry Pathfinders to the site of a planned raid, hence worrying about them as well. All the while not knowing for sure where he was, or what he was about to land on! Cautiously, he descended through that dark, swirling cocoon of thick dust. "The brown-out was so bad, and the field we just happened to land on, was plowed so deep, I thought the aircraft was gonna roll. We had actually slammed down onto the ground, planting that Black Hawk so hard, the googles snapped off my helmet!" Kline vividly remembered.

Thankfully, the helicopter did not roll, nor was it damaged. Most important of all, there were no injuries to anyone on board. The thick dust began to clear, and with it, the soldiers he carried were able to conduct their raid and complete the mission. Looking back on that close call, Kline remembered he'd had "some pretty sporty landings out there!" On this one, he remembered thinking to himself, if not outloud: "Thank God!"

Although he would go on to accumulate nearly 400 combat missions in the air, it was tense moments like that one, recalled Kline, that made him the most concerned while piloting in a war zone. "Most of the time that I was the most nervous, was not so much concern about getting shot. It was the actual flying portion. I almost had a mid-air in Iraq."

In that instance, at night, Kline and another unit ship, were both lifting-off from different operational LZ's (landing zones). Dark as it was, he was coming out of a big dust cloud, adding to the tension. "And the moment I did, I saw (the other aircraft's) left position-light right in front of me. It seemed like our blades where within a foot of each other. That could've been a really bad mid-air." Kline immediately radioed the other helicopter, being flown by a pilot he knew, and said, "Was that you? Holy cow"!! Upon recent inquiry, the Colonel did admit that "cow" was likely not the exact word he used to describe that far-too-close brush with severe injury, more likely, death!

Colonel John Kline was born in Orange County, California in 1970. With his dad, a 25-year Marine Corps officer (and former long-time U.S. Congressman from Minnesota), Kline experienced the "military brat" lifestyle, living "all over" growing up. Not only was his dad military, but his stepmom, too, serving 20-years as an Army nurse-officer. And his two Grandfathers distinguished themselves in the military, as well. One as a WW II (Normandy) Army First-Lieutenant, the other, a Navy pilot in WW II, retiring as a LCDR after, among other assignments, commanding the airfield at Corpus Christi, TX. Given that impressive lineage of service, the response to his choice of a military career might be expected: "It's what I knew as a boy growing up in a military family. Like father, like son."

Kline earned his undergraduate degree at Shippensburg University in Pennsylvania (small, close-knit school. I loved it.") and was commissioned a Second Lieutenant in Army Field Artillery. Following the F.A. Officer Basic Course, his first assignment was with the 4th Infantry Division, Fort Carson, CO. Three-years later, he transitioned from artillery to Army aviation, completing the rotary-wing qualification course (UH-60) at Fort Rucker, AL. With follow-on assignments, principally in Germany and at Fort Campbell, KY, Kline rose to command assignments, before, during, and following his combat deployments. In Iraq, he was the Executive Officer for the 5th battalion of the 101st CAB. During his two Afghanistan tours, he served, first, as Deputy Brigade Commander (101st CAB), and later as the Commander of Task Force Eagle Assault (5-101st CAB).

Colonel Kline graduated from the Army's Command & General Staff College (Fort Leavenworth, KS), and, in addition, has earned two master's degrees (Central Michigan University/Administration) and (Air War College/Strategic Studies). He has been awarded, among other distinctions: 2 Legions of Merit, 4 Bronze Star Medals, 4 Air Medals, 3 Meritorious Service Medals, 4 Army Commendation Medals, plus earning the Air Assault, Combat Action, Master Army Aviator, and Parachutist Badges.

Colonel Kline is married with two children. He served as Brigade Commander, 3rd Infantry Division Combat Aviation Brigade, Hunter Army Airfield, Savannah, GA, from 2013–2015. He deployed for his fourth combat tour in October 2015.

*Operation Swarmer, Iraq 2006. Major John Kline (4th from right/back row), Executive Officer, 5th Battalion, 101st Aviation Regiment, during brief shut down, a group photo with other crews, then the mission continued.*

John Kline vividly remembers his 2005-2006 OPERATION IRAQI FREEDOM experience, deployed with the 101st Airborne Division's Combat Aviation Brigade (CAB), as a time when he viewed Iraq as "the wild west, except that we didn't have to worry about getting shot at a lot." His first huge, combined mission there, was named OPERATION SWARMER, with Black Hawk and Chinook air assaults under that banner carried out continuously for eight-days straight. Huge because it needed to be, since they were stationed along with the entire Division at COB (Combat Operating Base) Speicher, near Tikrit in the north, with the 101st CAB supporting 101st Airborne soldiers on the ground. They had been given the task of clearing out, both insurgents and weapons, across a hefty amount of open desert terrain, stretching from Baghdad all the way up to Kirkuk.

To help better position this timeframe in the mind of the reader, amidst the hectic, on-going operational tempo in Iraq, OPERATION SWARMER occurred at about the same time that the Sunni uprising was taking place, and the Golden Mosque was blown up. "I flew over the Mosque site the next day, as I always did, looking for it as a landmark, since you couldn't miss it. Only now, it was completely gone!" As stated, with that deadly internal Sunni-Shia conflict well underway, likely directly connected with large-scale civil unrest, Kline remembers that "while flying around, there were times when we spotted decapitated heads in the desert, and beds of pick-up trucks covered in blood. Given the location, we were pretty sure that blood wasn't from slaughtering sheep!"

*Iraq, 2006*
*COB Speicher near Tikrit.*
*Major John Kline and his*
*crew following a mission.*

With SWARMER's initial sweep goals completed, next in Kline's Iraq battlefield memory came OPERATION EAGLE WATCH, an on-going mission, lasting about eight-months, developed and designed to take advantage of the complacency that had developed, among insurgents, regarding unaccompanied, apparently-routine, Black Hawk flights.

"The enemy saw two aircraft taking off, perhaps, ten, twenty times a day from these operating bases, and, with them, we were generally doing "battlefield circulation," just flying people around. But anytime we started to go up with a couple of Apaches, the militants knew we were probably coming to a village near them, preparing to land, with guys in the back who were going to get out and start kicking in doors and those sorts of things," said Kline.

So, to help disguise intent, EAGLE WATCH employed a different tactic, sending out just two Black Hawks, alone, giving the appearance of simply flying another transport mission. In actuality, each two-Black-Hawk-flight in this operation carried a total of twenty light-infantrymen, a medic and an interpreter. No question, these flights, unaccompanied, required an extra dose of courage from pilot and crew, but, again, at this point in the Iraqi conflict, ground fire was less of a concern. It wouldn't always be so, but in the earlier stages of OIF, thankfully, it was. This new tactic of swooping in with just the two Black Hawks landing near villages, off-loading infantry, and grabbing suspects of interest to gain hoped-for intelligence, "took off like wildfire," recalls Kline.

One prime example. EAGLE WATCH flight crews were supporting a battalion up North, southwest of Kirkuk. The ground guys were looking for a militant nicknamed 'Turkey,' whom Kline recalls was described as "a mid-level thug, who made and planted IED's." But every time our infantry soldiers went to get this 'Turkey' guy, it was a "dry hole." They couldn't find him. And so, Kline's team met with our infantry leaders at the FOB (Forward Operating Base) and told them what he felt they could do to help. "You tell us what's going on in your neighborhood. We've got two Black Hawks and twenty-two 'shooters' (Pathfinders) of our own." The first place the infantry commanders wanted Kline to look, was a certain village, in a certain suspected house, leads that were given with a frustrated, witsend look, as if to say, 'Hey, you'll never find this guy.'

Kline told them he'd give it a shot. He took off, and located the village, just about a three-minute flight from the FOB. "While we probably just looked like two Black Hawks flying to Kirkuk, we suddenly banked, landed in this guy's backyard, and all of our pathfinders came off with their weapons." Quickly, over the sound of the rotor blades, the interpreter yelled, in Arabic, to the first guy to appear in the doorway: 'Where's 'Turkey' '? Without hesitating, the man answered that he was right there inside the house!!

"Two or three minutes later," said Kline, "our soldiers emerged with a guy blindfolded, his hands zip-tied, the normal way captives were then prepared for transport." In no time, they'd gotten the much sought-after 'Turkey.' "So, we scooped him up, took off, and flew back to the Forward Operating Base. Total time we were gone was, I think, probably 15-to-20 minutes, and we said to the infantry battalion staff back at their TOC (tactical operations center), who predicted he'd never find him: "Here's 'Turkey,' " recalls Kline with justifiable satisfaction, given the infantry's prior negative prediction! To which the battalion operations officer replied, 'Holy (Bad Word)!' This terrific success proved the definite worth of the aerial tactic, conceived by Lieutenant-Colonel Don Galli and then-Major John Kline. It also proved that it's possible, out in the wild, for a Black Hawk to uncover and serve-up a Turkey!

And that tactic remained very successful, whenever needed, with succeeding EAGLE WATCH missions. Two solo Black Hawks, a combined twenty-two infantry guys in the back, with the task of going from village to village throughout that very large territory, and with surprise landings, removing anti-government fighters from them, preferably alive, as information sources. At the campaign's end, there were more than 100 air/ground missions under the over-all OPERATION EAGLE WATCH banner, making it a huge, lengthy, and effective air-assault campaign.

Along with intelligence gathering and rounding up any bad guys found, early air-assisted efforts in Iraq were, perhaps, primarily aimed at stopping the insertion of those destructive and deadly IED's, and especially preventing their detonation. At that time, our military was actually using civilian contractors, flying thermal-camera-equipped, real-time-video-gathering, Cessna aircraft to 'scan' the ground environment, with sightings of interest relayed verbally to the assault helicopters (the capability for direct video to aircraft cockpits would come later). Those small planes were able to fly up high and virtually in silence (compared with the helicopters), to avoid tipping off village observers. "When we fly low, we're flying fast, we're flying helicopters, so we're loud, we're in your face kind of thing," said Kline. Teaming up with the out-of-sight-out-of-mind air-scan aircraft brought our assault pilots closer to an almost stealth-like advantage, allowing for a faster approach, putting troops on the ground, more quickly, effectively cutting down insurgent reaction time.

Within a couple of weeks, Kline remembers getting a call from overhead that they'd observed a guy with a video camera, outside a village, planting IED's along a road the military traveled. When finished, he then hurried into an open field, dressed like, and hiding among, the typical sheepherders, except, recalls Kline, strangely, the suspect was wearing brightly colored tennis shoes! Kline's Black Hawk came in, and the infantry Pathfinders went to work. Searching through the herders, they found the guy with the unusual shoes, patted him down, found the detonator on him for those IED's, and, with his video camera, carted him off, preventing the planned blast and, also, the likely video of the destruction he'd hoped to create. After that take-down, an EOD (explosive ordinance disposal) team was called in, shutting the road down, while they safely detonated the IED's.

That task of trying to stop crippling IED explosions would continue, indefinitely, became the next assignment for Kline's team, zeroing in on a main U.S. military supply route (MSR Tampa) heading up northwest of Tikrit. Infantry units from the 101st Airborne Division were routinely having to deal with IED's along that route, and other transit points, at the rate of ten or more positioned and/or detonated daily! Naturally, commanders wanted it stopped.

What they thought might be happening was, throughout part of the terrain, a series of 'wadies' (big ditches or ravines) ran alongside the road. They felt the insurgents must be planting their bombs, then backing off to the wadi to hide, while watching for traffic, then blowing-up their desired target. The infantry felt that adding an air-assault component to the effort would make it more effective. An infantry combat logistic patrol was planning to head up that road, so the timing for the combined mission seemed right.

The operational planners at FOB Summeral said to Kline: "Here's what we'd like to do. We've got air-scan flying at about ten-thousand-feet, and they're going to be watching us (i.e., that combat logistic patrol) as they drive. And we'd like to have you and your two Black Hawks off-set about seven miles, with your Pathfinders on-board, and just hold out there in orbit." As the ambush scenario elements meshed, at least in theory, Kline wasn't certain all the pieces would actuality come together as envisioned, but he was eager "to give it a shot," as they worked to synchronize the timing of all the land and air components.

He remembers the operational conditions being far from ideal. "It was the middle of the night, with no illumination, super dark out, and very dusty." Regardless, he set out for the rendezvous location, and was in orbit there for about five-minutes, when he and his crew saw a huge flash on the horizon. It was quickly confirmed that the patrol had, indeed, just hit an IED on that road. Information was relayed to the air-scan pilots overhead, who replied that their thermal camera was showing a "hot spot." They noted three individuals hiding, as suspected, in a wadi.

Kline's aircraft moved in toward the given coordinate, but due to the darkness, they were unable to spot the insurgents. Then air-scan radioed that the suspects were running, and at that point, were actually just about below his helicopter. Almost immediately, his crew chief yelled that he'd spotted them. Kline landed his Black Hawk, and several Pathfinders, chasing after the three, tackled them, catching them red-handed with a detonator. Now in custody, they were blindfolded, wrists bound, and taken along for questioning.

In the dead of night, shooting the running insurgents wasn't an option, said Kline, since at that point their identity was unknown (e.g., could have been kids), and they weren't shooting at our soldiers (unarmed when taken down). Remembering that, as always, the primary goal of such captures was to interrogate to gain human intelligence. "And several success stories were like that one" recalled

Kline, "catching trigger guys, but most always catching kinda small-scale dudes. No big financiers, no really big high-value targets, just a lot of guys, but regardless, we were using a fairly dynamic technique to get them."

Additionally, control measures were increased, such as a 10 PM curfew, along with "SNAP TCP's" (traffic control points), interdicting any vehicle that moved on those roads thereafter. "It became a real cat and mouse game. One that we weren't yet winning," remembers Kline.

Another of those EAGLE WATCH missions that still stood out so clearly in his mind, was the one where word had been received about the suspected location of a high-value target, a head suicide-bomb trainer, thought to be an older Iraqi gentleman who had some kind of a leg problem, likely walked with a limp. Kline and his colleagues went looking.

There were two Iraqi villages, about 800-meters apart, thought to be in-play. The two Black Hawks on the mission landed, one near each village, and sent their on-board troops in among the homes and residents to look for the suspected bomb trainer. But after a preliminary search, they came up empty. With the approach of the aircraft and the appearance of soldiers on the ground, some suspicious-looking activity, from a few of the presumed inhabitants, was, however, taking place. "We're now flying off-set," recalls Kline, "and we looked down at the village that's right below us, and we spot a gentleman in a 'black man-dress,' with a cane, exiting a mud hut about a kilometer away."

Wondering if that's their bomb-trainer suspect, Kline contacts his Pathfinder troops on the ground, letting them know he wants to pick them up and reposition them. As his soldiers began coming back out of the houses, so did two "local" guys who made their way to a truck and began driving away. "The typical standard (for insurgents) was to put their hazard lights on, drive 5-miles-an-hour, that way they'll (i.e., the Americans) think it's not us!" (vs. a more normal high speed get-away attempt).

Then Kline observed the old guy with the cane beginning to walk south, while that slow-moving, two-person vehicle headed north. Soon, yet another guy, this one in a "green man-dress," came out of a hut and started heading east, all departures happening at about the same time, obviously determined to divide rather than be conquered. Kline concluded that these movements were beyond suspicious. Coordinating with the other Black Hawk, both aircraft landed, picked-up their respective Pathfinders, and lifted off again to observe the movement of the departing suspects.

One of the infantry teams was told to go after the truck, on foot, with weapons out. Convinced the guy with the limp was their primary target, Kline determined that he and his flight crew, alone, would go after him. "So, we're flying in his direction, a really skinny guy, who's now running (with cane!), but we lose sight of him briefly, then we see him again. He's at the bottom of a gorge. He sees us come in and land. I told the crew chief, Specialist Decanio, to go get him. As he does, the suspected trainer with a limp, throws away his cane, and suddenly becomes Carl Lewis!" Apparently, a miracle cure!

It's obvious, very quickly, that the suspect is going to out-run Kline's crew chief. So, he lifts off, flies forward down this large gorge, and lands again, this time in front of the suspect. Kline then tells his other crew member, "there he is, go get him!" About the time the second soldier closed-in on the fleet-footed guy with the "limp," Specialist Decanio, still running, catches up and tackles him. "We tape him up, with duct tape, that's all we had (!), threw him in the aircraft, the crew gets back on-board. So now we've got one of them."

The task, then, was to go get the other guy, the one in the green 'man-dress.' So, Kline took off again, flying across the plateau and over the wadies, when they spotted the guy, down in one of them. Kline lands near-by, and again dispatches his crew chief to capture the green 'man-dress' wearer, this time heading from the aircraft with weapon in hand, just in case. He exits the aircraft, but in so doing, the magazine had fallen out of his weapon! Amidst the adrenalin rush, intent on the capture, he didn't realize it. Unaware of his potential vulnerability, the crew chief, nonetheless, captures the guy, pats him down, gets him into the aircraft, and duct-tapes him. Then the crew chief, again, hooks up to the on-board communications system to report. Now that he's back safely, Kline tells him, with a certain amusement: "Decanio, you did a great job! Now, go back out and get your magazine. It's sitting in front of the aircraft. You didn't have any bullets in your weapon!" A humorous situation, rare in combat, but only became chuckle-inducing after the fact, with the crew member unharmed.

Kline flew back to the village, delivering the two captives to the Pathfinders, as the team there had successfully stopped that vehicle trying to discreetly get away. Weapons and IED making materials were, in fact, in that truck, as with so many others stopped by our troops. So with all four captives (the two walking away and the two in the truck) now blind-folded ("bagged") and wrists tied, they were led back to the village. And that's when things took a turn. As the insurgents appeared, women came out of the houses and began to wail. It would, very soon, get worse. Much worse.

One of the two truck riders in custody, a very heavy-set gentleman, suddenly became ill, extremely ill. It was quickly apparent he was having a heart attack. Our guys on the ground did what they could to try to revive him, but within minutes, he was dead. This set off even more wailing from the women, as more villagers began to spill out of their houses and gather around our Pathfinder team, the three remaining captives, and the newly-deceased.

The team leader in the village radioed Kline to advise him of the situation, triggered by the, now, unexpectedly dead suspect. Determining the scene there was deteriorating, after checking for instruction from higher command, and receiving same, including, specifically, disposition of the deceased enemy insurgent, Kline ordered his troops to re-board his aircraft, with orders to bring along only the three living captives, and all the materials collected, for transport back to base. Once back on the ground, with prisoners transferred for interrogation, Kline inquired, once more, about the decision to leave the dead man back in the village, an instruction that had been either made or relayed, he recalls, by a junior officer. At that point, Major Kline got a valuable tip from his battalion commander, should the deceased-captive scenario ever occur again: "Next time you have a guy die out there, bring him back, too. We can leverage that a bit." Reason being, the religion of those enemy fighters (and all others of that faith, of course) requires them to be buried within twenty-four hours after death, meaning the deceased could be used by our team to "barter" with the family, trading the body for information that would hopefully be useful to America's air/ground efforts.

Overall, throughout his deployments to both Iraq and Afghanistan, the mission-role executed repeatedly by Kline and his sister Black Hawk crews, could be summed up as follows: "Our piece was mostly putting guys in, taking them out, putting them up on mountain tops, taking them out." Would he then stick around, orbiting in place? "Yep, for a little while we would, should there be a need for immediate extraction or casualty evacuation. If someone gets hurt, I can pull them out of there, but, usually again, we'd fly off-set for just a little while, and we'd then leave our Apaches on-station to continue the fight."

Kline well-remembers that this was the time in Iraq when ground-based threats to aircraft were minimal. Early-on in Allied military involvement, the Iraqi militants concentrated their offensive efforts, as previously described, mostly on planting and detonating IED's (unseen they hoped), those traumatic explosive devices that caused such destruction to America's fighting force, in lives, limbs, and equipment. Another preferred weapon in the militant arsenal was the use of suicide bombers, individuals or vehicles, who chose, or were coerced into blowing themselves up around allied military bases or members.

Afghanistan, he quickly learned, was a totally different story, thinking back to his three Middle East deployments, the latter two to Afghanistan, following that first one to Iraq. The Iraqi militants were more "cowardly," said Kline, while the Taliban in Afghanistan would openly "fight back;" they were "ruthless."

"When we got to Afghanistan, that's when the shooting (at our helicopters) started and it became really dangerous," he said. "When I was a brigade XO (executive officer), with the 101st CAB (OPERATION ENDURING FREEDOM / 2008), while I don't remember the exact numbers, at that point, we were approaching a hundred aircraft shot down, shot up, or forced to deviate their mission, due to being engaged."

As one example of the battlefield severity he and other aviators now faced, Kline clearly remembers two incidents involving Chinooks in flight. With the first one, an RPG had hit the rotor blade, while flying in mountainous terrain, forcing an emergency landing, coming down directly on a rural mud hut! In the second incident, ground fire made eighteen entry/exit holes along the aircraft, with a full load of soldiers in the back. Somehow, that Chinook wasn't shot down, making it safely back to base. And here's the amazing, make that miraculous, part. Despite being shot down and shot up, there were no injuries on either aircraft!

And then there was his own experience piloting a Black Hawk, when, shortly after lifting off, an enemy mortar round struck and exploded at a spot right beneath his aircraft, the exact troop pick-up point from which they had just exited. The happenstance of timing had saved lives. While military pilot expertise is a critical component in combat, there's an old saying with which Kline would no doubt agree: Sometimes it's better to be lucky than good!

Clearly, this was a totally different environment from the one he had faced and flown through during OPERATION IRAQI FREEDOM, two years earlier. As brigade XO, a "re-education" briefing for his pilots, who had previously served in Iraq, was definitely in order. "Afghanistan is NOT Iraq! That's the first thing we'd tell them," Kline said. "The Taliban are not Al-Qaeda or any other insurgent groups you faced there. It's a totally different fight."

*Afghanistan 2010. LTC John Kline, with his Black Hawk, prepares for a mission.*

As mentioned earlier, Kline vividly remembered that the Taliban would fight back. "No longer those Iraqi fighters who'd take pot shots, or snipe at you, and then run-away kind of thing." In Afghanistan, if flying near a Taliban-controlled village, they'll move to positions where they can get down and put a heavy volume of fire on you. And they'd just harass some of our FOB's (forward operating base) every single day, like clockwork, lobbing off mortars, along with gun fire."

He remembers that his particular FOB at the time, "Wolverine," had more indirect fire (i.e., mortar rounds) hitting it than any other of our forward-operating installations. How best then to counter the danger? "We started putting up all these concrete barriers everywhere. We began flying a lot of local patrols. And we started doing a lot of ground patrols with our Pathfinders." Kline's description of the reoccurring daily and potentially deadly pounding they faced, made it all seem somewhat like the repetition portrayed in the movie, "Groundhog Day," only with people and equipment being targeted and shot at, abruptly ending any Hollywood similarity!

"We knew exactly which village the Taliban fighters were in" he said. "We'd occupy it one day, and Taliban, the next six! And the village elders would tell our soldiers: 'We have no guns. When you come to our town, we do what you tell us. And then we do what the Taliban tells us, too. And they're here more than you are!' "The frustration for our soldiers there was that we simply didn't have enough troops to take and hold all of those villages, leaving the Taliban free to come back in, a frustrating process of swapping control, time and time again, and always at a cost to our military in casualties.

It had become quite obvious, that when the Taliban wanted to move around, to set explosives or establish shooting positions, their preferred transportation of choice was the motorcycle. They were everywhere, outside and within the villages. And it had long been determined that anytime two young men were spotted on a motorcycle, no question, they were Taliban.

So, the countermeasure became interdicting as many of the two-wheelers as possible. "We started taking them out. We couldn't shoot them (rules of engagement did not yet permit), so we'd stop them," said Kline. "We'd put ten 30-mm explosive rounds in front of them from an Apache, to get them to either ditch the motorcycle or stop and put their hands up. Then land next to them, in a Black Hawk with Pathfinders, to see what's going on."

All too often, of course, the soldiers would find blasting caps and all kinds of IED-making materials in those saddlebags. Sometimes, the motorcyclists wouldn't stop, despite the rather clear attention-getting threat that a hovering Apache can provide. "They'd go right around the shooting, ditch the motorcycle, and run into a mosque, because they had a safe haven there." In those instances, the Pathfinder infantrymen would examine the abandoned motorcycle, gather up any bomb-making materials found for later analysis, then destroy the cycle, and head back to base.

After those continuing efforts to cut down Taliban motorcycle traffic, the next set of missions for Kline and his group involved trafficking of a different kind. They were tasked to do vehicle interdiction out in Afghanistan's Red Desert region (area south of Kandahar extending to the Pakistan border). Called the Red Desert because the heavier-grained sand is actually red. The problem there called for stopping and inspecting Pakistani vehicles, in particular, when it became known that Taliban personnel were carrying on an active, high-volume barter exchange, trading weapons and equipment (destined for the Taliban) in exchange for Afghan opium.

So, Taliban members were constantly traveling the main road between the Red Desert and the Pakistan border. There was also active trafficking between the two entities in gemstones, especially up in Afghanistan's Northern Provinces, since that region had heavy deposits of emeralds, sapphires, and other mined jewels, again an attractive trading commodity for obtaining Pakistani-supplied weaponry, motorcycles, etc. Those illicit commerce chains, both opium and gems, needed to be broken, especially so on the Taliban end, to cut down on their access to fresh supplies, arms and otherwise.

Kline's group focus would be in that southern, Red Desert region, where the opium trade was the strongest. "So, we started finding their clandestine labs and destroying them", he said. But by then, U.S. forces couldn't do those operations unilaterally, since that territory was assigned to the British, meaning that our aircraft worked cooperatively with their SAS (Special Air Services) commandos and other elite Brit units, much like the "Eagle Watch" operations in Iraq.

The key difference was that the operational techniques had, by now, moved well away from the former Black Hawk-Air Scan combination, replaced by far greater sophistication in just about three years' time. "The Brits now had the capability to see what other aircraft sensors could observe through their leader's monocle (a part of his glasses)," recalled Kline, "so that he could actually see, real time, the full-motion video that was being transmitted from either a manned or unmanned system." Kline's team then listened to satellite communication, as an aircraft orbiting overhead picked-up 'ground moving target indicators' (GMTI), meaning anything metal that was moving down there: cars, trucks, motorcycles, whatever."

What followed were continually updated grid locations, including direction, and even the speed at which the vehicle was traveling. Kline continued to detail the process: "So, identified first as moving metal, that information was picked-up by our crew in flight, as we finalized our plans to interdict the enemy suspect's route." The formula: Land near the road, dispatch the Brit team, apprehend the vehicle, then collect evidence and prisoners. That tactic proved very successful for a while, until the insurgents figured out what was going on. Recalls Kline: "It's good that we started landing a little bit farther from these vehicles we stopped, because I remember we had one detonate. The guys all ran out of the vehicle, and they blew it up. And we said, yeah, it's not a good idea to be landing the aircraft anywhere close!"

If the Taliban bailed, as they did in this case, the British SAS troops would try to run the enemy down on foot. Often, depending on the number of insurgents, that would take some time. Meantime, Kline's Black Hawk would temporarily continue to orbit down low, until fuel became an issue, since they were then way out in the desert with no access to a refueling source. So, he would land and pull one engine back to idle to conserve. "You didn't want to shut down an engine out there, because if you couldn't crank it again, you could be stuck. And sometimes, on missions like this one," he remembered, "we were fairly close to Iran."

If the Brits were taking an unusually long time to complete their objective, both capturing bad guys and removing any bomb-making materials from the vehicle, Kline would need to radio them to wrap up their work as soon as possible, when his fuel-state was approaching critical. The elite troops would then return, most often radioing their higher command for permission to "Blaze the Charlie" (i.e., enemy vehicle) before lifting off. It usually was. Wrapping it in detonating cord, they would blow-up the vehicle, re-board the Black Hawk, and take off, concluding the mission. Failing an ability to detonate from the ground, Kline remembers that the vehicle could, instead, be blown back to metal fragments by the compelling 30-mm fire from Apaches in the vicinity!

Unlike earlier days in Iraq, a typical mission-set in Afghanistan would now be two Black Hawks and two Apaches, or depending on the ground component needed, it might be an added Chinook, two Black Hawks and two Apaches. And also, unlike an earlier time in Iraq, large open desert missions had given way to much smaller, more concentrated ones. In this new scenario, with far more hostile and aggressive Taliban fighters, American aviators were forced to operate, in and around, what he termed "crazy terrain."

Kline recalled what a typical mission might look like in the northern regions of Afghanistan: "Let's say you have a village that's got about twenty mud huts in it," recalls Kline. "And the village is typically located at the very bottom of a deep valley, with gigantic mountains and big foothills all around it. Generally, the Black Hawks would come in, while the Apaches started looking at what the heck was going on down below. You've got a UAV that's on-station watching, 12-hours prior, checking out everything. You've got a bunch of intel guys checking signal communications, trying to figure out, and confirm, who's down there. But we don't want to tip our hand with Apaches loitering overhead, because then everybody starts to put their sandals on, put a fresh magazine in their AK-47's, and it gets real kinetic, real fast!" To try to prevent that reaction on the ground, as much as possible, the Apaches would either do an off-set orbit, or trail well behind the Black Hawks.

Another difference, now, from those days in Iraq, was where the Black Hawks would actually land. It used to be right next to a building in the target village (landing to the "X"), remembered Kline. But no longer. "We started to get shot at a lot, so we added about 300-meters to the LZ (landing to the "Y"), because that's their max effective range (the enemy AK-47's). So, we'd land out here, or even on the other side of a terrain feature (e.g., a hill), but still close, and our infantry guys would work their way into the villages."

On other occasions, with a mission far more covert, Kline outlined the differing procedures: "We'd fly off-set (Black Hawks) and, at the appointed time, come in high-up and put only about five soldiers on a mountain top (number of troops limited by the greater engine strain of high-altitude flight), placing just that 'support-by-fire' team up there (snipers and Javelin missile systems), with the Chinook carrying the main force (about 30), landing out of gun-fire range, somewhere down near the village itself. So, you'd have the supporting fires up-top, and the Apaches, too, would be working the area. Then the infantry force, on the ground, would start moving in and clearing the village out." Stacked up above them would usually be an Air Force AC-130 gun ship, maybe a flight of F-16's, or a flight of A-10's, or Navy F-18's. "So if things got dicey for the infantry, if they needed fire-power, either our Apaches could call it in, or the infantry could, since they would have an Air Force JTAC with them on the ground." As for the length of these kinds of missions, they would typically run anywhere from "4-hours to 36-hours," said Kline, "although with special operators, it's maybe two hours max. They go in, hit the target, and they're out."

With insertion of conventional Army infantry troops, their ground missions typically went longer, as they would "root around" looking for insurgents, or for safety's sake (aircraft, crew, and team in the back), insertions may even be delayed, totally, until first light, in order to mitigate the accidental risk, which for helicopters in Afghanistan, especially, was really high.

Recalled Kline: "There were a couple of catastrophic crashes (e.g., Black Hawk with Navy SEALS on-board in 2010), others from trying to land at nine or ten-thousand feet and just run out of power, that kind of thing. Browning out, rolling Chinooks; the flying environment over there is just hellacious. Altitudes are crazy, and you still have the heat, still have the dust. Add those latter components to the high altitudes, and now your engines are starved for power. You just can't do it. Actually, you can, but you need to train your butt off," noted Kline.

In order to do that, to better master the life/death demands of high-altitude flying requirement, the Army started sending its pilots to Colorado to gain the recommended techniques, and the necessary proficiency, principally through lessons-learned "from our Night Stalker brethren." As a result of this specialized training, along with the rigors of on-going piloting skills enhancement, and, importantly, the seasoning accumulated through actual battlefield missions flown, "I would argue right now," said Kline, "that within the ranks of the conventional Army, our aviators are as good as they've ever been. Now, Night Stalkers still lead the way. And we've always sent the very best out of our ranks to the Night Stalkers, as it should be." As a veteran pilot, and a Combat Aviation Brigade commander, no question, Colonel Kline has the earned, well-deserved credibility to make that assessment, one he offers with genuine pride in the Army's aviation community.

But he does fear that pilots in conventional Army aviation may lose that combat-experience edge, with the inevitable departure from the day-after-day demands, and flying savvy, imposed by the prolonged Iraq, and especially, the Afghanistan conflicts. And that can only be mitigated by lessons learned passed down from more senior, oft-deployed aviators to the junior ones, now with limited or no deployment experience. Couple that with constant flying in real-world training scenarios, the latter, of course, can only happen, in the author's opinion, if Congress provides a realistic level of funding to allow for both the necessary flight time, and state-of-the-art aviation equipment, in order to generate and maintain, as much as possible, that life-preserving combat-flight edge.

Looking back on two other Afghanistan missions, then-Lieutenant-Colonel Kline and his team (two Black Hawks and two Apaches), were dispatched to villages where Taliban militants had been fighting troops of the Afghan Army. Kline landed near one village and, this time, off-loaded a Special Forces team tasked to head into town to gain information from the locals. No sooner had they started, than the SF guys realized they had become the target of a hidden sniper, fortunately one whose aim, initially, wasn't particularly good!

*Mission ready, LTC John Kline
"Strapping on a Black Hawk."*

Kline and his crew lifted off to take a look around to see if, from above, they could spot the sniper who was threatening the mission of our SF soldiers on the ground. Before long, on a nearby hillside, the Apaches noticed an earth-tone sheet that appeared to have someone sitting under it, that sheet color chosen specifically because it was hard to detect with the naked eye. And because it helped dissipate the heat of the day, it made the shooter harder to detect via aerial infrared thermal sights, as well.

The Apaches then requested clearance to engage that, so he thought, hidden sniper. Once green-lighted, they promptly fired a ten-round burst of attention-getting 30-mm rounds directly into that sheet. To make certain the sniper was out of business, Kline landed and inserted a couple of SF guys who approached cautiously to check. They found a radio, sniper rifle and expended shell casings. And that Taliban shooter was quite dead. So much so, recalled Kline, that "the sniper's severed scalp actually blew off the hill from my rotor wash!" It's just a darn good idea never to attract the business-end of an Apache!

In a somewhat similar instance of troops off-loaded to follow-up on the target of air-to-ground fire, two of our Kiowa Warrior helicopters had spotted a Taliban IED triggerman in the process of digging a hide-hole next to his motorcycle, his presence a clear threat to our troops. With clearance, they commence firing on the insurgent. To double-check on post-fire status, Kline landed his Black Hawk near-by and dispatched a couple of on-board Navy SEAL's. They approached, as always, with caution, and good thing. Unlike the sniper described above, despite live-fire from the Kiowa's, this IED-detonation guy was somehow still alive, and armed with a supply of grenades, making him a continuing threat. But he'd be a threat no more, as the SEAL's efficiently parted him from his motorcycle, his grenades, and this earth. No report was filed on the injury-status of the motorcycle.

Looking back on his over two-decades of Army experience, when asked how he would define courage, Colonel Kline responded: "It's doing the mission or action you wouldn't normally want to do, because of the existing conditions or obvious danger, but you go ahead and do it anyway." On another of his many Afghan missions, this one on his first OPERATION ENDURING FREEDOM deployment (2008), Kline recalled the courage of those from his group, up-close, under fire.

Two Black Hawks and a Chinook had arrived, and landed, at a pre-arranged spot, to pick-up U.S. forces. No sooner had the Black Hawks landed, than "RPG's start whizzing across their noses, behind the tails, hitting the ground all around them," remembered Kline. At that point, the three pilots are on the radio, together realizing they're in real danger, with the urgent need to get out of there. Just then, an RPG hit one of the landed Black Hawks broadside. But with good fortune, under the circumstances. "I think it was just a stage-one detonation. There are two stages with an RPG-7. Had the second one detonated, it probably would've been catastrophic," he said.

But even though lives on-board the aircraft were spared by the limited detonation, there was still some serious damage, as the shell had punctured the Black Hawk's fuel cell located toward the back of the helicopter. "Now, they know they're in trouble. The left door gunner's got shrapnel in his legs, and the crew chief radios the pilot that, not only was that fuel cell hit, but now it's on fire!" recalls Kline.

With the whole row of back seats already burning, but with the smile of good fortune again, there was apparently little likelihood of an explosion. Realizing that he must quickly reposition his aircraft to avoid further hits, the pilot manages to lift off and move it just about 600-meters distant, enough to get it out of the direct 'kill zone.'

He landed and performed an emergency shutdown, while the remaining crew exited the aircraft. But despite the repositioning, the

Taliban troops, sensing blood in the water, began to converge on the downed Black Hawk, determined to complete the kill. With the enemy approaching, the pilot pulls his injured door-gunner onto the ground, lies on top to protect him, and with his flight helmet still on, takes his M-4 and starts picking-off the advancing enemy troops.

Meanwhile, the Chinook pilot, seeing what's going on, alertly lifts off, and lands in between the downed Black Hawk and the approaching Taliban troops. The handful of Special Forces soldiers, still on-board the Chinook, peel out the back and "with their M-240-Bravo machine guns, just start mowing them down," related Kline. After the crew members from the badly burned Black Hawk had destroyed any remaining sensitive equipment and materials on their damaged aircraft, those crew members were then waved onto the Chinook, and with all the SF guys back on-board, the two remaining aircraft flew out of there.

The Black Hawk command pilot, who protected his crewmate and fired on those kill-hungry Taliban, with effect and great courage, was later awarded the Silver Star. The Chinook pilots, bravely facing enemy fire to effectively block the Black Hawk, along with a definite rapid-fire assist from those Special Forces soldiers on-board, received the Distinguished Flying Cross for their outcome-changing reaction.

Apart from any specific mission, from all that he had witnessed, Colonel Kline had great, comprehensive praise for the courageous, unceasing rescue actions of the Army medevac crews, operating day and night, throughout the combat zones. There were countless memories of heroic actions by these crews. One in particular stood out in his mind from Afghanistan. Going right into a "super-hot" LZ, where nine soldiers had already been killed. "Flying into that whole mess up in RC East, under a hail of bullets, saving lives. They had two or three guys that were urgent-surgical in the back of their aircraft, and they only had so many hands, and they're tying everybody off, pushing fluids, doing the best they can, and trying to resuscitate them. There were hundreds of those stories," remembered Kline.

When asked to recall what missions or actions, over the course of his three combat deployments had caused him to draw deep-down on all of his strength and training, perhaps not surprisingly, the Colonel's response was all of them, whether actually doing the flying in both conflicts, or serving as the decision-maker on the ground as a commander in Afghanistan.

"It was the myriad of decisions that you're making on a daily basis, some of which are life and death. That's what takes the mental strength, day after day, mission after mission. Are we going to go up and get those SEALs who are on a mountain-top at 9,500-feet, when the weather forecast has just rolled in, it's terrible, and we can see them from the UAV (real-time video) and they're freezing their butts off? Or are we going to launch the one of the "gazillion" medevac missions when the weather is crap? Are we going to launch that medevac before the launch is even approved? Are we going to shoot the sniper on the hill or not? Are we going to shoot the guy we believe has just set off the IED, because we're afraid he's going to get away, and are we willing to accept the wrath if we haven't exactly met the rules of engagement? Am I going to let my Pathfinders go into the mosque and follow the guys we know are bad, and we know we're not supposed to go into a mosque?" Missions he either flew, or for which he had to make those hard-call decisions, all of which, for Kline, became the ultimate responsibility.

**UPDATE:** Colonel Kline was promoted to Brigadier General on June 2, 2019. John Kline pinned on his second star August 1, 2022. Currently, Major General Kline is the Commanding General of the United States Army Center for Initial Military Training (CIMT), U.S. Army Training & Doctrine Command, Joint Base Langley-Eustis, Virginia, and is responsible for annually transforming 130,000 civilian volunteers into Soldiers committed to defending our nation.

From his time in command of the 3rd Combat Aviation Brigade, 3rd I.D., at Hunter AAF in Savannah, Major General Kline has held several higher-level Command and Chief of Staff positions, both around the nation and abroad, prior to his current assignment. MG Kline served five combat tours, with over four years in Iraq and Afghanistan.

In 2021, he represented the U.S. Department of Defense conducting in-person meetings with the Taliban Political Commission (TPC) in Doha, Qatar. He found that task to be ironic. As he recalled: "I met with the TPC for six months to include during our withdrawal from

Afghanistan. After three previous deployments fighting them (Taliban), and trying to kill them, to find myself sitting across the table with their most senior leaders was interesting, to say the least!"

# CHAPTER 9

## From Air Force Navigator to Air Guard Wing Commander
## Georgia Air National Guard
## Colonel Robert S. Noren

Born in 1970, in Spangler, Pennsylvania, a small coal mining town, located between State College and Pittsburgh. He grew up in nearby Westover, PA, where he would then graduate from the second smallest public school in the state! So small, in fact, that it did not have a football team, a problem causing sports and physical fitness-minded Bob Noren (first love: football!) to have to find a work-around. He chose basketball, really liked it, and continually used it to keep physically fit through college and well beyond.

For his college experience, he chose Penn State University in State College, PA. It's one thing to choose a college. Another thing to afford it. Noren had worked two summers prior to Penn State enrollment, enough to pay for one semester. So, as he remembered, "I had to quickly figure out the whole cost issue."

Early in his freshman year, the unsolicited solution arrived in his dorm mailbox. It was a form letter suggesting that he consider applying for an ROTC scholarship! And that he did. Totally out of the blue had come his college cost solution and, with it, his first exposure to the United States military. He spoke with the contact person on campus, signed up for ROTC, competed for a scholarship, and thankfully, he received one! "So, I was able to stay in college. And from the very start, I owe where I am today to the financially-sustaining start I received from the military's ROTC program."

While it would end up taking him an additional half-year beyond the normal four to graduate, due to those required additional ROTC classes.

Noren majored in engineering and ended up graduating (December 1992) in the Top 25% of his class. He was commissioned directly into the Air Force. "The funny thing about the whole commissioning thing, he remembered, was, when I began my freshman year (1988), there were 164 fellow ROTC cadets. Four and a half years later, due to the national military drawdown (late '80's to early '90's), I graduated with only 33! I was so very thankful that I was one of those chosen to stay."

Once on active duty with the Air Force, Noren wanted to become a pilot, but his eyesight was not up to pilot training standards. He was told that, instead, he could become a navigator. At that time, he was at Davis Monthan Air Force Base working as an engineer for a while before moving on to navigator school. That call came and, with it, time to take another physical. But his eyes were still an issue. Referred to a Colonel, who was also an eye doctor, this ranking officer, sympathetic to Noren's navigator desires, determined that with glasses, his vision was correctable to fit within Air Force standards. Thanks to an understanding eye doctor Colonel, who just happened to be there on temporary duty, Noren was able to begin his long-awaited navigator training and his actual Air Force career. As Noren learned and vividly remembered: "Sometimes we have to look at the bigger picture. Do we have to follow the letter of the law, or do we have to do the right thing. Sometimes it's the little things, and I try to pass that lesson on."

Next stop: Air Force navigator school at Randolph Air Force Base in San Antonio, Texas. His desired career path finally and fully underway, Noren finished first in his class! "Frankly, said he, "I never thought I could do that. But I was so afraid of failing that I worked my tail off!" Other than quiet bragging rights and feeling really good about his class achievement, 'order of merit' for graduates then applied to the choice of stationing assignment. Noren's first-place class achievement enabled him to be the first to select his first post-school aircraft and stationing assignment, from among those listed. As he remembers, "In 1995, I got my dream job! As a lieutenant, you're not even supposed to ever get this one: Air Force Special Operations at Hurlburt Field, Florida!" And there he was starting his military career for real, but as a lieutenant surrounded by captains. "A lot of what happens in the military is luck and timing." And that certainly proved to be true with this highly-sought-after very first assignment in Special Ops.

Noren would be flying and navigating in the MC-130 aircraft. The "M," said Noren, stands for multi-mission with virtually all of the mission assignments in this very special American Air Force combat-transport aircraft, which typically and understandably, flies at night, often traveling behind enemy lines, and most often on single-ship missions. "We did clandestine flights, dropping off our most elite soldiers to go do the work they were assigned to do, all the while the enemy (hopefully!) doesn't even know we're there."

With that reality in mind, early on, Noren discovered the level of secrecy that would be required by his upcoming missions. "I signed into my unit's Director of Operations who was a junior lieutenant colonel. He offered his congratulations for being at Randolph. Then, happy talk over, he surprised me when he said: 'Watch CNN today to learn where you'll be tomorrow.' "What a great line," thought Noren, "I've got to remember that one! But when the colonel said it, in Noren's brand new position, it always proved to be true. It took me a few weeks to fully understand that prophecy. You watch the news daily to see what's going on in the world, and then to see (and know) where you're going to be sent the next day! It was all very exciting, since as a young guy, here I would be going to some of the same places that people couldn't even point to on a map, because there was some uprising, or Americans were being held hostage, something like that. We had to plan a mission to ensure that, with our Special Operators being flown in, that the enemy would for sure be taken out!"

Noren remembered flying with several different Special Operations units, and what a privilege it always was for him. "I got to work with some of the best of the best, and it humbled me every time I saw, up close on our flights, who they were and all that they were so skillfully able to do. All of them, very, very special people."

Now an experienced navigator, Noren remained at Hurlburt Field in Florida, from 1995 to 1998, during which time he was promoted to captain and was deployed to Kuwait for two-to-three months, still flying with, and guiding, the MC-130. Right before deploying, getting all his stuff together for the overseas trip, a long-time unit secretary, "an older civilian lady, who was like a Mom to all of us," came to Noren in the midst of packing, holding an application in her hand, that had come down from the Wing, with the competitive opportunity for younger Captains to be assigned to the Pentagon, and while working a couple of jobs there, he was told the officer selected would also be able to earn a master's degree at George Washington University." This wonderful, seasoned civilian employee told Noren that she thought "he'd be great" for the assignment and handed him the one application she'd been given. Busy getting ready to deploy, Noren recalled only spending about two hours on the application, due to his need to get back to packing, figuring what's the use, as he was just about to head into a heavy combat zone with that chance, ever the chance, and ever on your mind, that this could possibly be his last flight.

Then one mid-afternoon, somewhat later, while preparing for his next night mission flight, his Squadron Commander "suddenly came walking into our tent, and announced to me that, when we do rotate out of here (Kuwait), you're on the first plane to leave. I asked why is that, sir? So, it seems, he said, that you have a job lined up back home, Captain. Well, that news was like, holy crap, you never really know what all is going on." Turns out it was that Pentagon assignment application that he had so hastily filled out, and, unknown to him, that dear civilian secretary back at Hurlburt Field had pulled information and put all of the necessary packet submission paperwork together for him. "She was a real sweetheart," recalled Noren. And on top of that news, he soon learned that his wife back in the states was pregnant!

When he returned from Kuwait, after a short leave to check on his dear wife, he had to attend a seven-week squadron officer's school, as the initial part of his Pentagon tour selection. Then, at some point amidst it all, he and his wife packed up and made the move to the D.C. area for the next two very busy years, for both of them!

Captain Noren worked at the Pentagon days, and then went to the university at night. He was in D.C. from 1998 until the spring of 2000. He recalls his time at the Pentagon was both interesting and valuable. "It was such a great experience, because there were so few Captains working there. And because of the program we were in (with other selected Captains from around the nation), we were allowed to go anywhere and sit in on almost any meeting. It became such a broadening experience. Being able to better understand how the smartest men and women in the DoD think was just incredible for me. It was an unbelievably hectic time, with day and night work/study expectations. "But at the end of it all," recalled Noren, "I came out of it a lot more well-rounded. And along with that benefit, the military had paid for my master's degree!"

It was now time to return to the operational side. At that time, the upgraded version of the MC-130's ('Talon Two') were located in the United Kingdom, Japan, and at Hurlburt Field. Since he was told that the manning levels for this newer version were then low overseas, he ought to plan on assignment to either England or Japan. That likelihood in mind, Noren sold his still-owned home near Hurlburt in Florida. Then, about three-months prior to completing his D.C. staff position, he was told: "Congratulations, you've got a 'Talon Two' assignment back to Hurlburt Field!"

Initially, Noren privately indicated his disappointment, but he would soon come to realize that it was actually a blessing. "Once I got back there," he remembered, "I realized that since I hadn't flown in two years, I was behind my peers. So, I volunteered for every available flight, every exercise, every time the airplanes were going off station to practice with the Army "customers." I wanted to be a part of it again, because I had a lot of ground to make up, and quickly."

In his particular squadron, there was a mission calling for the Rangers on board to assault and seize an airfield. This flight included between three and seven aircraft. In such a formation, there was an assigned Lead Crew, which dictated everything about the flight and everyone else then simply follows. "That crew was chosen for a reason. And that reason, perhaps obvious, was because they were really good," said Noren.

Then in mid-summer of 2001, with the group just returning from a major exercise out in Nevada, one of the Lead pilots and the Lead navigator decided to leave the Air Force for civilian life. Then, on top of those departure, and totally out of the blue, then 9/11 happened! At that particular time, Noren was actually out in Albuquerque, New Mexico attending an instructor school for aircrews (typically eight members per aircraft), designed to help experienced crew members to more effectively teach others coming into the unit. With the shock of 9/11 sinking in, Noren was ordered to take his required instructor check ride and fly immediately back to Hurlburt. Once checked in, he looked at the list of crew names and assignments for the eight or ten aircraft currently there at Hurlburt. "So, I looked up to where my name was on the charts and saw that I had been put on Crew One. The Lead Crew. Lead Crew navigator!"

Noren was quick to admit that he'd never led a formation or a field seizure before. He was told: "After following for two years, this is going to be your first one to lead. You already know what right looks like! So, this is yours to plan and yours to lead. And now we're going to go punch the Taliban and al Qaeda in the mouth."

*Taking Off.*
*Captain Bob Noren*
*U.S. Air Force*

On October 6, 2001, remembered Noren, "we flew over to a location outside Afghanistan. There we met up with the elite Army "users," and spent several days planning a mission. And again, my first! It would be a double seizure, using both a fixed-wing side and a rotary-wing side. We assaulted a remote desert airstrip and seized it, which allowed our follow-on forces to come in. And, even today, it still amazes me. That was the actual event. Not a practice run. That was the real thing! I had gone straight into "hit" night.

I'll never forget it. It was October 19th at 1935 Zulu. That was the time we dropped the first American paratroopers into Afghanistan, and the very first troopers to go in came out of my airplane!" Along with Noren's lead aircraft dropping in our elite troops, along with the other MC-130's, ten minutes prior to that collective drop, a fixed-wing AC-130 gunship flew in with them to 'soften up' the enemy numbers on the ground, and doubtless their resolve, as well! Meanwhile, the rotary-wing aircraft were simultaneously seizing another objective 70-miles away from ours.

But just prior to his description of the planned drop from his aircraft, Noren was asked if he thought that the gunships had done a comprehensive enough job prior, so as to avoid any concern about the vulnerability of his own aircraft going in for the troop drop? With this inquiry at a key point in his narrative, it was obvious that the mental recollection of that very critical moment in time, a live or die moment, had triggered a definite, observable, twinge of emotion in Noren. Said he: "I remember sitting in the navigator position, when the Air Mission Commander (in charge of all of the assaulting aircraft), who was then in continuing radio contact with the accompanying AC-130 gunships flying high above, received a radio alert that, regardless of the softening efforts prior, there now appeared to be active enemy fighters down below, and they seemed to be right under our intended flight path, information we received within just one-minute of our jumpers coming out!

I looked at our flight path coordinates and reinforced the mission commander's concern. Our current intended approach path would in fact, have put us directly overhead of still-active enemy fighters, with the distinct possibility of taking fire, possibility fatal to our aircraft, to our elite 'passengers,' and those that followed our lead."

At that point, as Noren remembered, those higher-altitude AC-130's let loose on the apparent still active enemy forces, with their impressive fire. Needing some extra time to accomplish that, no doubt, live-saving action, in order to protect his own troop-carrying formation, Noren "had to drag his group of MC-130's around and away from the active enemy danger, and lose five-minutes from the intended drop time, but far better, because we all said to ourselves that we're not going to do anything to knowingly jeopardize the mission of these guys in that back of our airplanes.

But when we finally did bring the formation in, down below there was a smoking hole that hadn't been there before, courtesy of the AC-130 gunships. Those terrific gunships would rotate back and forth to a circling tanker in the distance, so that there was always at least one active, fueled, AC-130 aircraft on-station, circling and providing constant protective coverage from above. Noren's team was then more safely able to head in at the correct altitude to complete the planned drop of our elite American forces, the first of the Afghan conflict. "That was the most memorable mission from my time in Afghanistan. We remained there for about 90 days, then returned home to Hurlburt. A few months later, I flew over again, for a total of about three, 60-to-90 days each in Afghanistan. Then we started preparing for the invasion of Iraq in December of 2002."

*Captain Noren's Crew*

"If you remember the First Gulf War," recalled Noren, "there was a very, very long air campaign for over 100 days, before they allowed the ground forces to move in. This mission was going to be different. A very short air campaign, and then the Army and Marines would go from Kuwait to Bosnia and on into Baghdad. Before Noren's team could go in, President Bush gave Saddam Hussein 48-hours to leave the country. While his departure clock was running, Noren's team received approval for a secretive very low-level flight into Iraq, heading toward the northwest side of Baghdad, then landing on, and seizing, a desert strip of land there, while dropping off our "Army customers."

Flying out, Noren's team then returned to base, loaded several vehicles onboard, and then returned to that same spot near Baghdad. Off-loading equipment and troops, "we then wanted to take off and get back out, but we were stopped from doing so."

Turned out the reason was the presence of U.S. Navy Aegis Cruisers, out in the Red Sea, launching Tomahawk missiles that were flying right over Noren's location. Needless to say, not a good time for them to take off! "We could see those missiles," he remembered, "flying over us at about a thousand feet, and looking very much like telephone poles crossing in the sky!" Although he was running low on gas, the air controllers, understandably, wouldn't let him take off. "We were burning gas, starting to run low, sitting on the ground in Afghanistan, while those things were flying overhead. It was surreal. Since there was nothing we could do, we might as well just enjoy the show!"

Noren was over there in the fight for three or four months. He flew back to home base at Hurlburt Field in May 2003. "At that point, I was actually thinking about getting out of the Air Force." Obviously, a big decision. But at that point, he wasn't sure. So, he took on one more assignment at Hurlburt. And he also deployed one more time to be the air liaison officer for one of the Special Forces groups in Afghanistan. While Noren considered that job to have been terrific experience, once he returned to Hurlburt, he realized just how mentally and physically tired he had become with the seemingly non-stop demands and deployments. As he remembered: "When we weren't deploying, we were flying hard practicing. It's kind of a young person's game." He had been promoted to Major in 2003, and even though his career was going really well, he couldn't get past the reality that he was just plain tired!

So, in the Fall of 2004, with two small children, and wife Bobbi's full support, "we left the Air Force and moved back to Pennsylvania. I took some time off. Including working some in my dad's construction company building houses. I just tried doing things that were totally different for a change." As it happened, his desire for something completely different lasted for about a year!

One day, Bob Noren just happened to remember a conversation he'd had, while he was still on active duty, with a retired two-star general who was then working as an advisor/consultant for a company, as retired military officers often did. The general gave him his business card, offering to help him find employment, if he ever could use his assistance. Noren kept that particular card. Smart thing that he did. After that year of construction work in Pennsylvania, that gnawing feeling of wanting to do something different came roaring back.

Somewhere toward the end of 2005 or the start of 2006, retrieving the general's card that he'd somehow managed to save, Bob Noren gave that retired two-star a call. Happily, the general remembered him. He was by then working with a defense contractor at MacDill Air Force Base (Tampa, Florida), Special Operations Command. The general told Noren to give him time to make a few phone calls. Sometimes in life, magic happens. This was one of those times. "Within a week or two at most, I got a call from somebody asking me to come down to Tampa. They had a job to offer me, which would be working in the Special Operations Headquarters as a civilian contractor!" Noren accepted that offer and before long, along with his family, he moved to the Tampa area. Turns out "it was one of the best jobs I'd had outside of the military. I really, really enjoyed it." And that's primarily because he was then working with some of the same military veterans he had deployed with three or four years earlier! But because of that continuing work proximity, before long, he realized that he was beginning to miss the military itself. "I began to miss wearing the uniform. So, I talked with a couple of buddies, and with my wife, and one discussion led to another.

Then one day out of the blue, he got a call from someone with a Georgia National Guard squadron located in Brunswick, Georgia. The caller indicated that he'd heard that Noren might be interested in re-joining. He was invited to come up and visit with them about the possibility. Best yet, he could even give the Brunswick Guard a try on a part-time basis, so that he could maintain his full-time job with Special Operations at MacDill. This was in the 2007–2008-time frame. Noren drove up to Brunswick, interviewed, and not surprisingly, given his prior service qualifications and experience, the Brunswick Guard officials liked what they saw, and he was invited to join the Georgia Guard unit there. With his own desires for a return to service, and with the all-important encouragement from his wife, Bobbi, to go back into the military, after about a four-year break from service, Bob Noren signed on with the Guard as a part-timer (165th ASOS), while maintaining his full-time position at MacDill AFB. The change in his mental state, with this decision to re-enter service, was probably summed up best by his wife who observed that he had now gotten his "mojo" back!

Remembered Noren: "I enjoyed it. I love it. It was a great change of pace. And within three-years of being with the Brunswick unit, still part-time, the command asked me to go full-time with the Guard, taking on the Brunswick Director of Operations position (2011). The first time they asked, I said no, because I really liked my part-time set-up. Then about a year later, they asked me again. Fearing they might not ask me a third time, I finally accepted. I left my job at MacDill and moved up to Brunswick, 'geo-batching' for about a year, with my family still down in Tampa." Then, when the Brunswick 165th Air Support Operations Squadron was moved up to Savannah, Noren and his family moved to Savannah (2012-1213).
"I was very fortunate. I served for about four years as ASOS Director of Operations (first at Brunswick, then Savannah), when they asked me to become the squadron's commander. I was the commander for about three years, and I was getting ready to retire (2019). Then they asked me if I would come over to the Wing headquarters (Savannah). Still ready to retire, I said okay, I'll serve in whatever capacity you want me to," recalled Noren.

So, from 2019-2020, he served as the 165th Wing Chief of Staff. At that time, serving in full-time command on an every-other-year basis, the Wing Commander was, that particular rotational year, a part-timer (flying full-time with an airline). Noren's primary responsibility, then, was to be sure the commander had all the information from the unit so that he could make the right operations and policy decisions. Then, "since they were short of 165th navigators, they asked me to come fly. I agreed and went back to flying after a 15-year break! I thought I'd do that for a little bit longer. Little did I know that, about a year after that, they promoted me to Colonel (LTC in 2012 / COL 2020)."

Noren then volunteered for a duty assignment overseas (2021). It was about a five-month tour to be the commander of the 385th Air Expeditionary Group, a C-17 unit in Qatar, the lone Guard person, with active-duty personnel working for him. For the first couple of months, as Noren remembered, it was "kind of a vanilla mission" (flying airlift throughout the 19 countries within Central Command). Then, without warning, it suddenly turned into something incredibly serious.

*Major Bob Noren calling in an Air Strike.*

And that was when President Biden announced (April 18th) that the U.S. "would be going to zero in Afghanistan"!

So, recalled Noren, "we went from seven C-17's and I think maybe eleven aircrews to the point where, within two-weeks, we suddenly had roughly twenty C-17's and about thirty-four air crews! And our main job was to just keep shuttling stuff out of Afghanistan, as quickly as possible, and bring it back to Qatar, Kuwait, or any other place. We went from kind of a sideshow business to the main effort in about a week!" It was his unit, then, that was in the thick of the on-going evacuation of materials and service members, to include the eventual concluding, punishing, and painful (the deaths of U.S. service members) chaos that Americans got to witness here at home via the on-going network news coverage at the Kabel airport.

Prior to that chaotic conclusion, in June 2021, his promised five-month overseas assignment now up, he left Qatar for home, while things were still incredibly busy there. Two months later, back home with the 165th, he was asked to become Wing Commander! "Once again," said Noren, "that kind of surprised me, but I humbly accepted" the opportunity for overall unit command responsibility. He would go on to serve in that capacity for a total of twenty months.

*Colonel Noren accepting Georgia 165th Air National Guard Wing Command.*

Then, while preparing for a normal 165th Wing drill weekend in early February of 2022, Bob Noren got an unexpected phone call from the National Guard Bureau. They asked him "if he'd be interested in leading the C-130 tactical airlift effort to support the European deterrence initiative." In other words, using the 165th C-130's to assist with flying in supplies to help the Ukrainians, with the Russian invasion then underway. "And my answer wasn't just yeah, it was an enthusiastic heck yeah!" recalled Noren.

*The women who support Colonel Noren:*
*Mom, wife Bobbie, and daughter Katie.*

Plans for a regular drill weekend were quickly put aside. Guard airmen had no idea what was underway until they arrived, as usual, on Friday evening. "We just turned the schedule upside down," recalled Noren, "to go from a training mindset, to one of being prepared to get out the door. By Saturday, we already had over 100 names of unit volunteers who had talked with their civilian employers, spouses, and families, and were ready to leave early Monday to fly over to Europe! It was amazing to watch the Wing transform from a steady-state training base during a regularly scheduled weekend, to the hectic frenzy of rapid preparation to go and do what we had all signed up to do!"

"And all I did at that point was just be their coach. I just watched everybody and provided a little bit of feedback here and there. But it was really the men and women of the Wing that got everybody ready, and it was phenomenal to see. It was the highlight, because when they got over there, they were conducting some of the most important airlift missions to the Ukrainian forces, which were directly impacting the state of combat operations over there."

Those 165th Guardsmen were flying rockets, javelin missiles, and whatever else the Ukraine soldiers needed to defend themselves. Flying from large airfields in Europe on into some small Air Force airfields just <u>outside </u>of Ukrainian territory, to avoid upsetting, or over-stepping American diplomatic relations and requirements. Concluded Commander Noren: "They were flying in armaments as opposed to eggs and bacon! And the real highlight for me was to see this Wing transform itself into something that I knew it could be."

The time frame, again, was early February 2022. Those Ukraine-assist 165th volunteers returned in May. The well-earned and deserved welcome home gatherings for families and Guard colleagues wouldn't be able to last very long, because just two days after their return, the Wing would undergo a higher headquarters inspection! Said Noren: "We had to pivot quickly to an inspection, which lasted about seven days. And we ended up doing the best this Wing has ever done on a major command level inspection. When you hit back-to-back home runs (arms delivery to Ukraine + great high-level inspection), that's something to be mighty proud of. No question, those were two of the big highlights during my time in Wing Command. And best yet, they weren't something I did. They were something I witnessed." But clearly it was Bob Noren's clear direction and command leadership that enabled the Wing's Guardsmen to perform so well.

In his remaining days as Commander, Colonel Noren was asked about his thoughts on leadership. What in his view are the key attributes or qualities that define leadership effectiveness. "I would say this, and it may sound very basic, but before you do anything (as a command leader), you need to understand why you took the job. If you took the job because it's a promotion, or because it's a status symbol, that will in large part guide you in the decisions you make. But if you took the job to make the people around you better, who will in turn make the organization better, that in my view is the key to leadership effectiveness."

"So, every decision has to be grounded in, and guided by, making your people, and by extension, the organization better, because ultimately that's why we lead. You have to have that constant dialogue with yourself: if it's not right, don't do it! And one of the keys for me is that I don't ask my airmen to do anything that I haven't done or won't do myself. Since BG Steve Westgate served in command, I think I'm the only commander who stayed fully tactically qualified in our aircraft. To me, it was important to keep up with all of the requirements those flying have. That way, I know the degree of difficulty myself that, in turn, will be required of others. All the while continuing to serve the unit as a professional officer."

The conversation with Colonel Noren then moved from leadership to another very important command quality: character. He responded: "You've got to be true to what's right or the whole thing, your command and respect, will unravel in front of you. I remember something one of my past bosses told me. When you get into a position of leadership, there will be some who don't want you to have that job. Maybe it's only a few, but they will find every reason to try to make you stumble. So, you have to go the extra mile to do everything right.

For instance, anytime there's money or influence involved in anything we do, you have to make certain that, as the leader, you don't ever take advantage of your position. Here's an example: When we travel, we get a per diem for breakfast, lunch, and dinner. So, you get to the conference and lunch is provided. When you get back to base, on your travel voucher, you need to 'unselect' the lunch money. Even though it may only be 12 bucks! You have to make sure that in every little thing you do, you are above board, not just because it's the right thing to do, but because if you don't, inevitably, someone will notice. Another example would be the staff car provided for the commander's use. I don't drive it off base unless I'm headed to a sister base for a meeting, or some other official purpose, and there are usually a couple of Guard people riding along with me. I'd never use it to run personal errands! I simply don't ever want to give the impression that I'm using a government resource for my own purposes. Examples of perhaps little things, but they must always add up to doing the right thing overall."

With leadership and character in mind, we then moved to the question of whether Colonel Noren could recall an officer he'd served with who might best exemplify the key leadership and character traits discussed above. He responded: "I served under a lot of really good officers in the Special Operations world, all of whom were impactful. But the one who really stands out is Lieutenant Colonel Paul Havel, one of my commanders early on. He had a lot of what we would all call 'Havel-isms.' He had an endless number of great lines that would stick with you and really make you think. One that really stuck with me was: 'You never want to do anything to hurt the baby.' So, I'm thinking, what in the world does that mean? Well, he'd go on to explain. Said the LTC: This squadron you have is like a baby. Everybody wants to be around the baby. Everybody is a part of the baby. But the baby will spit up. It stays awake all night. It's cranky. It interrupts your schedule. It gets sick a lot. It's needy. But the baby is also something very special, so you've got to care for it. You've got to feed it. You've got to love it. You've got to comfort it. And you've always got to be there for it. Because the baby's gonna grow up and make you proud. So, whatever you do, he said, if you're a leader in the squadron, 'don't hurt the baby'! And the more LTC Havel kept talking about this 'baby' concept, the more sense it made.

There will always be those in the organization needing more attention; those who will screw up assignments, etc. But then at the very next turn, they come around, and it'll make you incredibly proud of what they've done. The lesson being: by your actions as the leader, don't ever do anything to hurt the unit. Don't do anything to 'hurt the baby.' I just thought that was a very good way to look at the ultimate value of one's organization, and in this case, the 165th Airlift Wing!" Bob Noren served with LTC Havel when the latter was the commander of the 165th Air Support Operations Squadron in Brunswick (GA).

And finally, he was asked about what made him the proudest during his time in command. Said Noren: "It's those moments when the airmen finally embody the commander's intent, then they move out confidently and start executing the mission so effectively, that they almost leave you in the dust!"

*Colonel Robert S. Noren*

And how about the support level for our military here shown by the Savannah community? His response: "This is a very special place. You can't go to the grocery store, or pump gas, or like I did today, stop for some lunch, without somebody taking the time to thank us for our service, to the point here I almost feel guilty because I feel like I didn't do enough to warrant it. You go to the St. Patrick's Day Parade in Savannah, or go to some other major community function, and the pride shown by area citizens for those now serving is sincere and truly appreciated by those of us in uniform. In a word, Savannah's support for our military members is phenomenal."

**UPDATE:** Colonel Robert S. Noren relinquished command of the 165th Airlift Wing, Georgia Air National Guard, in June 2023. He currently serves as the Deputy Director for Resources & Requirements NGB (A 5/8) at the National Guard Readiness Center at Andrews Airforce Base, near Washington, D.C.

# CHAPTER 10

## A Personal Tribute

## The U.S. Army Career of
## LTG Donald E. "Rosie" Rosenblum
## on the Occasion of His Memorial Service at
## Hunter Army Airfield

At son David Rosenblum's request, and his assistance, it's my honor to provide brief highlights of the exceptional Army career of Lieutenant-General Donald E. "Rosie" Rosenblum, who passed from our midst on September 6, 2022.

As most if you know, Rosie attended The Citadel in Charleston, where he was very active in student organizations and athletics, completing his degree in 1951, as a Distinguished Military Graduate, with commissioning as a Second Lieutenant in the United States Army.

With the Korean War then underway, in no time, Rosie found himself, there, leading a rifle platoon in combat. And from that time, came at least three stories he enjoyed telling, many times over.
On very likely, his first night out, keeping watch for the enemy, along with his platoon, Lieutenant Rosenblum was positioned along a short embankment. He soon became curious about a strange sound, going

on around him and overhead. He turned to his trusted Senior NCO, and asked if he knew what that sound was. "Yes, sir, replied the veteran Sergeant: "That sound is enemy fire. They're shooting at us, sir. Suggest you duck down." Sage advice, and combat reality lesson 1-0-1.

That NCO's name was First-Sergeant Joe Gomez, whom Rosie grew to really like, admire, and respect. So much so, that he faithfully kept in touch with him over the years. As his Army career progressed, each time Rosie received a higher promotion, he would drop a note to that favorite Sergeant, letting him know. And each time, said Rosie, Joe Gomez would respond with congratulations and the following: "Frankly, sir, I never thought you'd make it past Captain!" Rosie always enjoyed that sarcasm from his dear friend and would chuckle each time he told the story.

The final reference to his time in combat in Korea had to do with the weather. He mentioned, many times, that his time out in the Korean winter was the coldest he'd ever been. And whenever Savannah's winter-time temperatures would approach the 30's, it inevitably brought back memories of the extreme freezing cold of Korea, where he and his soldiers had apparently not been issued adequate winter clothing. Whenever he spoke of that extreme cold, colorful language followed.

Stateside assignments followed Korea, including two years with the 82$^{nd}$ Airborne Division at Fort Bragg. After time in Germany, eventually came, two combat tours in Vietnam, including command of an airborne infantry battalion. During his time with, or leading, various airborne units, he ended with about 60 career "jumps." Early in 1975, by-then-Brigadier-General Rosenblum, was assigned, from his position at the Pentagon, to go to Fort Stewart and Hunter Army Airfield to re-activate that post, prior to the planned arrival there of the 24$^{th}$ Infantry Division. As he often told us, Fort Stewart had only two or three habitable buildings, due to neglect from lengthy inactivation. Rosie rallied the limited on-post staff, and contractors, to repair, build-up, and make the installation suitable for an eventual full division.

With the project completed, well enough to begin taking on the sequential arrival of troops, Rosie then traveled back to the Pentagon, with what he hoped would be a simple request. He asked his boss if he could become the Deputy Commanding General of Fort Stewart/Hunter, that is, the second in command, since it was only the two-stars who got divisions and, at that point, Rosie had one. To Rosie's immediate surprise and disappointment, his boss leaned across the desk and said: NO!

In typical Rosie, stand-your-ground fashion, he asked why he was being turned down. The reason, said his boss, was because he was about to assign Rosie, not as the deputy, but as Fort Stewart and the 24th Infantry Division's overall Commanding General! The Number One slot! Rosie had no issue with that. His second star would follow. He had leap-frogged over existing Major-Generals to achieve that command, a clear indication of his record of achievement and the high esteem with which he was viewed by Army superiors. Major-General Rosenblum commanded Fort Stewart-Hunter, and the 24th for two-years. Following another command, in 1980, he became the Deputy Commander of both the 18th Airborne Corps and Fort Bragg. A year later, he was promoted to Lieutenant-General, and assigned to command the First United States Army, with overall training and readiness responsibility for three-hundred-thousand Army Reserve and National Guard forces east of the Mississippi, along with several thousand regular Army troops.

In between Rosie's operational tours, there were, of course, a number of staff assignments. For the sake of time, I've limited these highlights to the operational side. I do want to mention, however, that, between his two Vietnam deployments, he was chosen to attend the Army's War College, in Carlisle, Pennsylvania, a highly selective one-year posting, and a clear indication that the Army intended considerably more growth and responsibility for Rosie, which, as previously indicated, definitely followed.

Then, in 1984, after 33-years of exceptional combat and peacetime service, Lieutenant-General Donald E. Rosenblum chose to retire from the United States Army.

During his time commanding Fort Stewart & Hunter, Rosie had become very good friends with the legendary, influential, and, as it turns out, persuasive, Savannah Mayor, John Rousakis. When the Mayor caught wind of the General's pending-retirement, he contacted Rosie and didn't mince words in his message, which was to: "Get your, (let's say, 'rear end,') back here to Savannah!"

Still accustomed to taking orders, apparently even from one particular civilian Mayor, Rosie did, in fact, make the decision to come back to Savannah. He opened an office here, as most of you know, and ran his own very successful military-industrial consulting firm, for well over 20 years. In addition, he became and remained very active in civic affairs within the Savannah community, along with assisting his alma mater, The Citadel, in every way that he could.

Today, we honor, recognize, and most of all, remember a great American, one who became, as well, a distinguished military warrior and leader. An exceptional soldier, father, grandfather, friend, and patriot, who through grit, skill, character, determination, and professional achievement, rose to the very highest levels of military rank and command. He served our exceptional nation with duty and distinction, both at home and away, with unfailing commitment to our great Army, and to the cause of freedom.

In Rosie, we've lost a man of remarkable statue, a man of incredible accomplishment, and for so many of us here, a dear loyal friend. We will miss you deeply, Rosie.

And now, after your 33 years of dedicated Army service, permission, Rosie, to remain 'at ease' for all eternity, within the blessings, warmth, and peace of your heavenly home. Justly due, following a professional life, well done, dear faithful and distinguished American soldier.

# CHAPTER 11

## Esteemed Military Pilot and
## Flight Instructor Extraordinaire
## LTC Jack Scoggins, USAF (Ret.)

Air Force Lieutenant-Colonel (ret.) Jack Baxter Scoggins, born in 1931, was raised out in a country home, near Logansport, Louisiana, some 30-miles south of Shreveport. So far out into rural Louisiana, in fact, that in order to function, "daylight had to be piped-in to them," he recalled, which we'll take on faith was more a slice of sarcasm, than actual reality. Following high school, where he participated in JROTC, Scoggins attended Louisiana Technical University (Ruston, LA), graduating in 1954 with a degree in mechanical engineering, and as an ROTC cadet there, also received a reserve commission as a Second Lieutenant in the U.S. Air Force.

His desire was to secure an assignment in the Air Force appropriate to his engineering degree, but slots to his liking were not then available. His only other reasonable choice: pilot training. For the record, unlike oft-heard tales of men and women whose dream was to flying from a young age, that was not the dream or the goal of engineering-oriented, fresh new Lieutenant Scoggins. In fact, at that point in time, he had no thoughts at all of making the Air Force his career. But for the time being, his official orders, desired or not, were to pilot training, a sequence destined to develop within him, not only a military aviation career (a widely varied one, as we'll see), and a lifetime love of flying.

Scoggins followed the prescribed pilot training sequence back then, first to Columbus Air Force Base (Mississippi), where he learned the basics in Piper Cubs and T-6 trainers. Then onto Reese AFB

A Salute to American Patriot Warriors

(Lubbock, TX) to gain proficiency flying T-28's, and, back then, WW II-vintage B-25's. Final stop on the training sequence, heading forward, now, as he was, on a multi-engine path, in 1955, Scoggins was sent next to Donaldson AFB (Greenville, SC) to train on the aircraft he would then fly for the next 16-years, the comparatively-massive, certainly back then, C-124.

The C-124A Globemaster II entered Air Force inventory in 1951. It was a four-engine, long-range, heavy-lift, multi-role aircraft, flying with a crew of five (pilot/co-pilot/navigator/two flight engineers/loadmaster). When asked about handling this very large plane, Scoggins only memory of concern was the fact, that on final approach, the huge nose of the aircraft naturally came up, obstructing the vision pilots normally enjoyed in the moments before touch-down! All in all, he recalls that he "loved to fly" the C-124. It was fairly easy to fly, said he, "once you got the hang of it!"

Jack Scoggins literally flew the Globemaster around the globe and back. From flying cargo well into Northern Canada in support of America's defensive Distant Early Warning (DEW Line) site work, to missions throughout Europe supporting U.S. Air Force bases there, to Antarctica re-supplying scientists working at McMurdo Sound, to flying cargo through the Berlin Corridor, proving to the East Germans our capability and intent to remain steadfast. From time to time during the latter flights, Soviet MiGs would come up to "escort" these transport flights in and out of the corridor, certainly a "nice" gesture, though hardly welcome!

Then, in 1958, Scoggins began a three-year tour at an AF base in Japan. The principal mission was to bring in supplies, and especially fresh vegetables, to American troops stationed in South Korea, as well as transporting our military members to and from R-&-R breaks in Japan. During this time, events in that part of the world were re-igniting. On one such fight, the Scoggins crew was dispatched to Bangkok, Thailand to evacuate U.S. personnel from our embassy in Laos. Also during this period, Scoggins and another pilot were sent on a secretive mission to a Southeast Asia nation, assigned to travel the country looking at aircraft accommodation capacities of the airfields they located there. Important information ahead of the hostilities that would soon begin. It was during this mission-challenging time, stationed in Japan, that Jack Scoggins received his promotion to Captain.

In 1961, he was transferred from that pre-war part of the world to a completely different place and responsibility. For the next three-years, Captain Scoggins would serve on the ROTC instructional staff at Tulane University in New Orleans, preparing cadets for military service and, with the way of the world situation at that time, likely eventual deployment to yet another Far Eastern fight.

Following the ROTC assignment, Captain Scoggins returned to the air, spending two-years (1964-66) flying C-124's out of Hunter Field in Savannah, Georgia. His supply and support flights were, once again, worldwide, to include South Vietnam. Those roundtrips could extend to 3-weeks in duration, transiting from Hunter to California (Travis AFB) to Hawaii to Guam to the Philippines, and on into South Vietnam, then reverse course to home. It was during this stationing that Scoggins was sometimes involved with the transport of nuclear weapons, both as a training officer (the safe handling of such, air & ground), as well as piloting missions for delivery of such anywhere in the world they might be needed. Fortunately, the deployment and actual use of such weaponry apparently never came to pass during the Vietnam conflict.

After his Savannah tour, it was back to Japan at Tachikawa Air Base (west-side of Tokyo), a main cargo and troop carrier hub, for a second stationing there. Along with his flying duties, he also served as the Airlift Command Post Officer-In-Charge, with overall responsibility for transport aircraft loading, unloading, and getting all flights into and out of the base on-time. In Japan, he was attached to the 22nd Squadron, the last of the active-duty Air Force C-124 squadrons in the United States, as other airlift assets were making their way into America's transport inventory.

Over a 16-year period, Jack Scoggins had flown about 5,000-hours in that work-horse aircraft, in the latter years, to include supply flights from Japan (Tachikawa) to South Vietnam. On one such landing in DaNang, U.S. F-4's were flying close overhead, bombing enemy troops off the end of that same runway! Making it a much "hotter" and more challenging landing for Scoggins than he and his crew had normally experienced! Sometimes combat did come too close for comfort. After landing at Tachikawa after one mission, the maintenance crew called Scoggins later to let him know they had found two bullet holes in his C-124's flaps!

After all those years flying the venerable C-124, a decade-and-a-half and 5,000-hours worth, service life for now-Major Scoggins was about to change dramatically. As 1970 approached, he received new orders from the Air Force, sending him back to the states, first to survival school, then to helicopter training, and upon completion, on to stationing near Vietnam! Little about these orders sat well with Scoggins, except possibly the return to the states part! Regarding the Vietnam-area assignment, he recalls his first reaction was "Like hell you are!" But, as every service member knows, or will remember, orders are orders, and Major Scoggins packed for home.

Unexpectedly in all of this, he would be transitioning from a heavy airlift, multi-engine transport plane, to learning to fly military helicopters. As it turned out, his transition was not at all that unusual at the time. With our involvement in the effort to contain the Communist threat in Vietnam, attempting to protect the southern half of the country especially, the need for helicopter crews had become so great that U.S. pilots were being pulled from all fixed-wing aircraft, not only transport, but from bombers and fighters as well, for cross-training in piloting helicopters, mandated by the intensity of the growing war-time need.

So, Major Scoggins was by no means alone in this war-time-demand transitional phase. His first stop state-side was Fairchild AFB (Spokane, WA) for survival school, a critical step for those who would soon be flying in areas thick with ground fire. The course began with five days of POW indoctrination, typically followed by a 7-to-10-day stint in the woods learning to survive on the land. As luck would have it, and much to his delight, Scoggins was pulled from the school prior to the woods phase, because his helicopter training class was scheduled to begin. So rather than a week alone in the wilderness, he headed to Shepard AFB (Wichita Falls, TX) for a four-month introduction to the whole new world, for him, of flying military helicopters, beginning with the UH-1 (Huey), followed by the twin engine HH-3 (Jolly Green Giant).

Then it was off to Eglin AFB (Florida's western panhandle), where

he first encountered the much larger, more capable HH-53 (Super
Jolly Green Giant).

*HH-53*

Globalsecurity.org describes the Sikorsky HH-53 as "the first
helicopter specifically designed for combat search and rescue
operations. It was faster and had nearly triple the take-off weight of
the HH-3...larger, more heavily armed...with better overall
performance and hover capability, especially at altitude."

When asked about the difficulty transitioning from fixed wing to
helicopter flying, he recalled that "it was kinda rough." The hardest
thing he remembered having to do, initially, was to work out
controlling the landing on a precise spot. "You've got to descend and
then stop right over that landing spot, and for the first while, I had
the darndest time stopping that cotton-picking thing over the
spot!" As you might imagine, for a pilot with his experience and
determination. it wasn't long before he was putting those helicopters
down, from the beginning Huey to the Super Jolly Green, right where
they needed to be. Scoggins recalled that the HH-53 was actually
easier to fly than the Huey. Among other reasons, it had a lot more
power. "It was a very big helicopter, and I had just come out of a
very big airplane, so that's probably why I liked it so much."
Transitional flying training finished; Jack Scoggins was off to
Vietnam.

But before his final destination, there was a five-day stop in the
Philippines for jungle survival school. And on this round, it was time
to go into the woods (jungle, actually!). The objective was to stay out

there overnight without being spotted. Young local men were offered a reward of one pound of rice for every American they were able to uncover out there in the darkness, a handsome incentive in those days. So off Scoggins went into the jungle, positioning himself back off of a trail, convinced that, by sun-up, he'd remain undetected. He told himself "I'm gonna make it, I'm gonna make it." The next thing he remembered was feeling a hand on his shoulder belonging to a young Philippine lad, asking Scoggins for the "chit" that would prove he'd earned yet another valued pound of rice.   Later, Scoggins asked one of his buddies how they'd found him. As it turned out, no real mystery there. "You fell asleep. You were snoring like hell," said his buddy! To guard against that give-away in the future, should he need to be in hiding, once arriving in Vietnam, he promptly secured an ample supply of No-Doze tablets!

The year is 1970. Now in the Vietnam combat region, Scoggins was stationed at an American Air Force Base in Udorn, Thailand. It was from that location that he and his unit comrades would fly their combat search and rescue (SAR) missions, the principal purpose and responsibility of the Super Jolly Greens. Eight of the Jolly Greens were situated at four different locations, all on constant alert, with two of them on heightened alert forward, at a site in Laos. For the latter, on a rotating aircraft basis, two HH-53's would fly there each morning, land, monitor the radio, and wait for a rescue call. Near dusk, they would then fly as close as possible to that day's designated "strike zone," circling at a safe distance, in the event of a pilot down. Receiving no call, with night fully upon them, they would make the return flight to Udorn (there was no nighttime rescue capability at that time). Anytime two crews were scrambled for a rescue, two additional Jollys would fly to their alert location to cover for them.

A C-130 ("King Bird") was always airborne at altitude for each rescue mission, serving as the on-scene operations supervisor and communicator, both with those involved with the rescue effort and with our military authorities in Saigon (all rescue launches had to be authorized by Saigon).

Along with the on-scene supervising aircraft, a second C-130, serving as a tanker, also launched on every mission, ready to refuel the Super Jolly Greens, when needed, especially critical after an extended time-on-station, first while searching (and/or dropping fresh batteries to the downed pilot to maintain radio contact, if he was attempting to move to a safer location), and then during the hopefully rapid actual pilot rescue, the fresh fuel allowing for a safe flight back to the designated base. And speaking of the need to refuel in-flight, Jack Scoggins memory is crystal clear on his longest flight: 8-hours & 15-minutes, all without leaving the pilot's seat!

HH-53

Air Refueling with

C-130

Scoggins had originally been assigned to DaNang, but at that point in the war, there were no HH-53's at the facility in DaNang. The Super Jolly Greens, the helicopter he was trained and assigned to fly, were, however, at Udorn. And there's an important sidenote here. Jack Scoggins wife, an Air Force nurse, was currently assigned to Udorn. Scoggins got on the phone, and as luck would have it, the person he reached in personnel was an old friend. Scoggins explained about the availability of HH-53's, but not at DaNang, and, oh by the way, mentioned that his wife was already stationed at Udorn. The friend in personnel promised to look into his transfer request. Two weeks later, he got his revised orders to Udorn, Thailand! Sometimes things do work out right!

Major Scoggins (on the right) with his HH-53 crew.

Jack Scoggins would fly virtually all of his missions out of the U.S. base at Udorn. A few of those missions and experiences remain vivid in his memory. One such was the assignment to take three crews and two HH-53 aircraft to set up a small flight facility inside South Vietnam. Scoggins was the ranking officer with responsibility for getting this small, satellite airfield up and running with communications, crew quarters, etc. At one point during that process, the call came to scramble for a rescue. Typically, two aircraft would respond, one flying up above the other ("high bird"), for back-up, while the other ("low bird") searched the ground for our downed pilot.

Since Scoggins, as senior officer, was then elbows deep in the logistics of the operation, he turned his helicopter over to another pilot, and the two aircraft then lifted off to commence the search. Before long, word was radioed back that the "low bird," the aircraft Scoggins would have been flying, had been shot down, with all five lives aboard lost (pilot/co-pilot/engineer/two para-rescuers ("PJ's"). Sadly, Scoggins called back to his squadron commander at Udorn to report the tragic loss. He remembered being startled at first by his commander's initial response over the radio, when he said immediately: "Oh, thank goodness, Jack. Thank goodness it wasn't you!" The reason for his relief to learn that Scoggins had not been killed on that flight, in the very aircraft he would normally have flown, was not out of any lack of remorse for the loss of his squadron crew members, but rather, as Scoggins recalled, because his commander had dreaded the thought of having to go over to the base hospital at Udorn to break the news to Jack's wife face to face! Five courageous men had taken Jack Scoggins aircraft and mission response that day. A mission he normally would have flown. That fact, and the resulting loss of his comrades, still weighs heavily on him.

While finding and rescuing downed American pilots was the primary mission, occasionally there would be others. One such involved extracting several Laotian soldiers, in danger of being overrun by Communist forces, along with family members (and sometimes even family animals, the so-called "people and pigs" rescues!!) from a small airfield in Laos. This involved pulling a few of them on-board, then hovering to see how the helicopter was handling the weight. Then repeat until they had reached maximum weight under maximum power. In that particular instance, at the end of the runway, there was a big drop-off. So, recalled Scoggins: "We would begin our take off, then drop down into that valley, get our flying speed up, and fly out of there." He remembered that they had undertaken a couple of missions like that, but pilot recovery remained the number one priority.

A typical day for Scoggins, his crew, and others, would be flying from Udorn up to an allied airfield in Laos (or other airfields in the region) and spend the day standing alert, awaiting the call for a rescue flight. One of those subsequent rescue calls should never have occurred. It resulted from an inappropriate action (perhaps the kindest way to express it) on the part of an F-4 Phantom pilot. F-4's fly with two officers, a pilot and a rear-seater, known in the F-4 as the RIO (Radar Intercept Officer). This particular pilot, experienced and, thus, should have known better, was flying that day with a brand new RIO. The flight was over an active enemy combat area. And it was at night. As it was related to Scoggins from a post-rescue debrief, while in flight, the pilot asked the new RIO "if he'd ever seen ground fire." The RIO answered "No." To which the pilot replied "well, I'll show it to you!"

The F-4 was on a reconnaissance mission at the time. He had flown across a "hot" valley to help determine where the enemy troops might be concentrated. To "show" his new RIO enemy fire, the proved-to-be-foolish pilot then pulled up, "does sort of a tear-drop maneuver, and flies back down low, right back into that same valley, and gets shot down," related Scoggins. The two crew members ejected and parachuted to the ground.

"What hurt the most," recalled Scoggins, "was the next day, when we were there trying to get the two crew members out, we heard over the radio that the RIO had been found, shot and killed on the ground by the enemy." Scoggins learned that the pilot was alive and had made it to a cave. "We worked that mission for three days," he recalled. "He's in this cave, he says his back is hurt, and his radio is giving out as the batteries wore down. So we go in and drop batteries to him." The downed pilot was able to safely retrieve the new batteries and radio contact was re-established.

On the second day, said Scoggins, "every time we'd send someone in to pick him up, the rescue helicopter was shot at and had ended up full of bullet holes. So, the aircraft would have to pull off."  Finally, F-4's and A-1 Skyraiders ("Sandys") were sent into strafe and bomb the enemy combatants to try to free up the area for the rescue. And then, said Scoggins, "the HH-53's would try to go in again, and, again, they'd get clobbered, and would have to pull out." Clearly, the enemy had figured out that as soon as the F-4's and the "Sandys" stopped attacking, the rescue would be attempted, and so the ground fire would re-commence whenever the rescue helicopter re-appeared.

So the third day, the rescue team decided to try a different tactic. The pilot was hiding in a cave on the other side of a mountain, somewhat removed from the enemy fire. "So, we kept the F-4's and the "Sandys" bombing and strafing the entrenched enemy in the same location as before, while bringing the HH-53 in from a different direction." The "low bird" was able to safely low-hover now, and they lowered the "penetrator" with a para-rescue man to pick up the "injured" pilot. At that point, recalls Scoggins, "the pilot came out of that cave like he was doing the 100-year dash. Clearly, there was nothing wrong with his back!"

Nine helicopters had taken part in this rescue effort, and six of them were shot up so badly they struggled back to home station but couldn't fly again until extensive repairs were made. "One came back, and the landing gear was just dangling," he remembered, "so we ran to the barracks and got mattresses and laid them down on the runway, so the pilot could land on a pile of mattresses, without having the rotor come down and hit the concrete. That helicopter was leaking fuel and was full of bullet holes, but the crew was down safely. When it was all over, that common-sense-impaired pilot finally rescued, at tremendous expense to aircraft and great risk to crews (two members among the rescue crews did receive bullet wounds), Scoggins summed it up by simply saying: "He caused us a lot of trouble!!" Spoken with the composure of an experienced pilot, that is the understatement of the decade. And as Scoggins recalls, it wasn't long after, that some of the impacted Super Jolly Green pilots had, shall we say, an intense encounter with that particular F-4 pilot, making their point regarding his errant judgment one to be remembered!

The A-1 Skyraider, known to the rescue community as "Sandys," was a single-propeller- powered attack aircraft flown in Vietnam by both the U.S. Air Force and Navy. Key attributes were its favorable loiter time, and its ability to carry considerable weaponry (four 20-mm cannons, plus several hard points for rockets, bombs, etc.). The "Sandys" worked in close coordination with the Super Jolly Greens to suppress ground fire so that downed airmen rescues could hopefully be safely made. As with all combat aircraft in a war zone, it was dangerous work, requiring perhaps even more daring, due to the necessity of the Sandys flying down so low over enemy positions to be effective in ground fire suppression. According to Wikipedia, "The USAF lost 201 Skyraiders to all causes in Southeast Asia."

That by way of background for Scoggins other successful rescue effort resulted, this time, from pilot bravery, as was normal, not foolishness. Scoggins was flying his Super Jolly Green on a mission to retrieve a downed airman. As was most often the case, two HH-53's were assigned to this SAR call. On this day, Scoggins was flying the "high-bird," position, while the other was in the "low-bird" slot, with primary responsibility for locating and picking up the downed U.S. survivor. The "high-bird's" role is to observe, and it anything happens to the primary rescue Jolly Green, the high one's task is to come in to rescue that crew. With F-4s scanning and protecting the skies from up above, as usual it was the ever-present companions and guardians of the Jolly Greens, the "Sandys," that were flying in low, strafing any intruding enemy forces, doing all in their heavily armed power to clear the area for the pilot rescue. Watching the low-level ground-clearing work of these daring pilots, he recalled then witnessing the lead "Sandy" get hit by enemy fire. "He was hit badly," said Scoggins, "I mean, he was on fire!"

That "Sandy" pilot then pulled up into the air not far from where Scoggins, in his "high-bird" position was orbiting and observing. "So, we got in behind him and radioed: "Sandy-1, this is Jolly 87…. Punch out, we'll get you." Scoggins was very familiar with the area, as they had been working it for two days. "It was a hot area, real hot," he recalled. But rather than follow Scoggins advice to bail out, the "Sandy" pilot just pulled out and kept on flying, no doubt hoping to make it back to a friendly airfield. "By this time, the flames were past his cockpit," remembered Scoggins who flew to keep up with him.

So, he repeated his instruction to the pilot: "Sandy-1, this is Jolly 87, right behind you. The flames are past your cockpit!" The pilot responded, "Roger that," and just kept on going.   He wanted to get as far away as possible from that "hot" combat area.

"Well, by this time, the flames tore back behind his tail," said Scoggins. "We moved over to the side and did not get behind him anymore, because we didn't know whether he was going to blow up!" Scoggins radioed the pilot his flames observation again, assured him they'd rescue him, and once more told him to "punch out, punch out." Finally, at that point, the wingman of the endangered "Sandy" pilot, flying nearby with him, yelled into the radio: "Oh, damn it, Tim, get out !!!"

And with that from his nearby wingman, the "Sandy" pilot finally ejected. Scoggins watched as the pilot seat separated and the parachute came out, as the pilot headed for the ground. Complicating Scoggins' immediate rescue response was the fact that, after flying in that area for some time, previous to the "Sandy" being hit, they had just air refueled. "So, we were loaded, making us high and hot, so we could not hover with the load that we had. So we had to blow the auxiliary tanks to dump fuel."

Now, in addition to safely ejecting from an aircraft engulfed in flames, that "Sandy" pilot, with a simple round parachute, no guidance capability, had somehow managed to drift down and land on the top of a mountain in a cleared area! Having successfully dumped sufficient fuel, Scoggins' Jolly Green was then able to come to a hover just about the time the "Sandy" pilot hit the ground. The crew lowered the penetrator near the incredibly fortunate pilot and pulled him up to safety. "I don't think he was on the ground for even a minute," recalled Scoggins. Once on board, in response to questions about his condition, the pilot indicated he was fine, but hoped he hadn't scared the crew by shooting his gun on the way up into the helicopter. When asked if he was responding to ground fire, he told them, no, he just "wanted to shoot his pistol." Coming as close as he did to perishing in an aircraft likely about to explode, and then somehow landing in a perfect spot for a safe extraction, perhaps that pilot can be excused for firing off some adrenalin. Pilot retrieved, with fuel an issue, having dumped a share of it to allow for the pick-up hover, Scoggins then headed for the nearest allied airfield, which was DaNang in South Vietnam.

And one important footnote. The original downed pilot, the reason for that particular mission to begin with, was, in fact, rescued. The remaining "Sandys" successfully cleared the area of enemy fire, so that the "low-bird" Jolly Green could hover and make that survivor pick-up. A successful ending for two downed pilots, and their rescue crews, all the way around.

But there is an unusual, if not truly amazing coincidental, twist to this "Sandy" rescue story that occurred some thirty-years later. Then retired from the Air Force, Jack Scoggins had become a highly regarded commercial pilot and the owner of a flight school at the airport in Valdosta, Georgia (his wife, then still serving as a senior nurse officer, was assigned to Moody Air Force Base outside Valdosta). As a designated pilot examiner, Jack was required to take a flight-check every year from an FAA inspector to renew his certification. On this day, he had to fly up to Knoxville, Tennessee to take his annual glider flight-check from an inspector at the District Office there since, at the time, no one in the Atlanta office was qualified for glider certification testing.

Previous to his arrival in Knoxville, the FCC inspector had advised Jack that he had three exams to give, but there was only time for two. He told Jack that he'd give him his exam first, and then asked if Jack would, in turn, administer the test to the remaining candidate so that he, too, could obtain the commercial glider add-on to his commercial flight certificate.

Jack was happy to oblige and the flight-check for that particular candidate was successfully administered and completed. Following the exam, while he was finishing up the paperwork, the candidate happened to notice a Jolly Green patch on Jack's flight jacket. That triggered a conversation: Had each served in Vietnam, when were they there, etc. Come to find out they had been in the same vicinity at about the same time (1970), and the gentleman he had just glider-certified indicated in their conversation that he had actually been a "Sandy" pilot there. Letting him know that he had, of course, flown the HH-53 in combat SAR missions, Jack said: "We probably flew missions together and didn't even know it!"

With the "Sandy" connection established, Jack began to relate the saga, detailed above, of that determined "Sandy" pilot, his plane on fire, the eventual bailout, and his safe rescue by Jack and his crew. Jack then added the footnote about the pilot shooting his pistol on the way up the penetrator, just for the sake of doing it. To which the gentleman replied: "Sounds just like, Tim." So Jack asked if he had actually known that pilot. Without hesitating, the gentleman replied: "Who in the hell do you think was his wingman that day!" Incredible. After all those years, Jack had just successfully flight-checked, and engaged in conversation about their 1970 Vietnam service, with the actual wingman who had demanded that his "Sandy" squadron-mate eject, certainly helping to save his life. The adage "small world" doesn't even come close to an unbelievable coincidence like that.

Returning in time, now, back to 1970, at Scoggins' home station at Udorn, Thailand. As he vividly remembered: "One day, we go to our intelligence briefing, and I see nothing but red spots all over the briefing map… heavy gun emplacements, and SAM sites all over the place." Their tasking that day was to fly up to the western part of Laos, a frequent staging area to await rescue calls, a location known to the pilots as the 'Fish's Mouth', located west of Hanoi, North Vietnam, explaining the heavy defensive concentration ringing the area. Typically, they would fly to a holding area like that, then at about two-hours before dark, they would fly in even closer to the active air-combat area, so that, in the event of a rescue call, they could hopefully get it completed before total darkness overtook the area. As Scoggins mentally processed the intelligence briefing that particular morning, and what he and his crew would no doubt be facing that day in the air, with the continuing likelihood he'd be called in to make a rescue, he remembered saying to himself: "Well, Jack, today, you're not coming back." Facing the prospect of bad outcomes, that extreme apprehension in the face of dire circumstances for the Jolly Greens, would later be replaced by miraculous news. The Jollys were on-station, flying apart from the target area, but in the vicinity, there, as always, in the event of an aircraft shoot-down and the need for a survivor rescue.

Scoggins recalled that, on that day, for that operation, we sent up over 100 U.S. bomber and attack aircraft, in a collective attempt to blanket-neutralize that extremely "hot," heavily defended enemy zone. And it worked, and safely, beyond all expectation and belief. That heavily armed armada of U.S. aircraft succeeded in taking out <u>all</u> of the enemy's missile and anti-aircraft sites in and around that particular target area! Even more amazing, it was done with <u>no</u> American casualties. "Not one was shot down!" he remembered, with still to this day, a combined sense of relief, amazement, and awe, realizing the negatives for our pilots that might have been.

The scheduling of "high-bird" and "low-bird" assignments was done as equally as possible, within the reality of combat demands and conditions. By trying to balance the high and low positions, each of the Jolly Green pilots and crews faced the same level and frequency of exposure, especially, of course, for the "low-bird," which was the primary rescuer (and potential enemy target), unless it was hit, in which case the "high-bird" would drop down into the fight to take on, at that point, perhaps multiple rescues.

In everyday life, clear and accurate communication is important. In combat, it's crucial, often making the difference between survival and sacrifice. Scoggins learned of one such case, sadly the latter, involving the loss of a Jolly Green, prior to his arrival in theater. When the high and low Jolly Greens fly to, and determine, the rescue location, two sister aircraft elements accompany them to assist. The "Sandys" go in on low bombing and strafing runs to neutralize the site prior to the rescue attempt, while, secondly, high above, F-4's fly what is called the "MiG-Cap," patrolling the skies above the downed-pilot area to keep it clear of any enemy MiG or other airborne interference. On that particular day, the Jollys radioed to double-check that they did, in fact, have top-cover U.S. fighter protection in place (i.e., the MiG-Cap). The reply came back in the affirmative. Whether misinformation or misunderstanding, the reality was there was <u>no</u> MiG-Cap in place, at the time, during that ground rescue operation. Unbeknownst, then to the vulnerable "low-bird" Jolly Green, flying above it and unopposed, an enemy MiG fighter in the area spotted the "low bird" and shot it out of the sky, killing all crew members onboard. An unnecessary tragedy caused by a communication error, during the "fog of war." The assumption would be that other protective aircraft were rapidly dispatched to that rescue site, and the original downed pilot retrieved, but that eventuality eluded Scoggins' memory, at the time we spoke.

He does clearly remember a non-flying tragedy at the Udorn, Thailand air base, which turned out to be a very close call for him, personally. Out on a mission, a USAF F-4 had been shot up badly and was struggling to make it back to the base. Due to the damage sustained by the aircraft, on final approach, its hydraulics failed. The pilot and co-pilot were able to bail out and survive. But, sadly, the F-4 then crashed into the base Armed Forces Television Network building, killing everyone inside. The resulting fire also destroyed two nearby BOQ's, though fortunately, all personnel there were able to get out without further injury or fatality. The reason this event remained so vivid in Jack Scoggins' mind is that he had been in that very AFTN building just about 5-minutes before the crash! An extremely close call on the ground, let alone his ever-dangerous missions in the air.

During his Vietnam tour, Jack Scoggins would fly over 50 actual combat SAR missions, to include the rescue of two downed pilots, both of which efforts, in addition to saving their lives, earned Scoggins the Distinguished Flying Cross (two DFC awards overall). But despite the acclaim represented by these and other service awards Jack Scoggins received, he was always quick to remind that all missions flown were true <u>team</u> efforts.

At the conclusion of his one-year duty assignment to Thailand, Scoggins returned stateside to Forbes Air Force Base (Topeka, KS) where he was assigned to fly H-3 helicopters with the National Geodetic Survey, as the Wing's Maintenance Quality Control Officer. As the name indicates, under direction and guidance from NOAA, his wing's mission was all manner of survey work, primarily, of course, aeronautical. He recalls not being there very long, because, several months after his arrival, they closed the base. Scoggins maintains the closure was in no part due to his performance!

Then, in 1972, the Air Force sent Major Scoggins to Patrick Air Force Base (Satellite Beach, FL). After completing the required water survival training course, he was back flying the HH-53 helicopter, this time preparing for the possible need for space capsule recovery in the ocean (both shallow and deep-water), or on land, should a problem occur for the astronauts at, or near, the launch site.

*LTC Jack Scoggins
at Patrick Air Force Base, Florida,
preparing for a possible astronaut recovery
flight.*

"At each launch pad, they had 120-degrees around it, where there were landing spots," he related. "If a fire broke out or if they had to abort and get out of the capsule real quick, they would grab onto a roller attachment and slide down an escape cable to the ground, and we'd be sitting down there waiting for them. Depending on the problem, it would dictate which landing site we'd use." That was one scenario, he remembered. The other would trigger, just before launch, or immediately after it. Said Scoggins: "If they didn't have a chance to get out of the capsule and onto that escape cable, they would fire a rocket on the capsule. That would take them up into the air, a parachute would come open, and then they would land either on the land or in the water. We were trained for three different recoveries: land, shallow water, and deep water." As Scoggins recalled: "About four-minutes after launch, they were out of our range. At that point, our job was done." If the astronauts had to land farther out in the ocean, that became the responsibility of another recovery team. Scoggins expressed understandable relief that, on his watch, he and his crew did not have to make any rescues. Good news for him. Great news for the participating astronauts! Turns out, there was some great news for Major Scoggins, as well. While he was on the Air Force assignment with the NASA astronaut rescue mission, he received his promotion to Lieutenant-Colonel.

In 1974, after 21-years of distinguished military service, distinguished literally with the awarding of two Distinguished Flying Crosses for his combat rescues, LTC Jack Scoggins decided to retire from the United States Air Force. When his officer wife's AF nursing assignment took them to Moody Air Force Base in Valdosta,

Georgia, they decided to establish their residence there in his military retirement.

But it didn't take Jack long to visit and become a fixture at the Valdosta Airport! Having obtained his flight instructor certification in New Orleans, back during his ROTC teaching assignment at Tulane, Jack took over instructing duties there from a departing teacher, working first for Holland Flying Service at the airport, before joining with two partners to form Air Valdosta, a flight training firm he managed, while also continuing to instruct students. After three years, Jack returned to flight instruction on his own, as well as piloting corporate aircraft for Valdosta area companies. In addition to basic flight instruction, since 1977, he had also been an FAA-Designated Pilot Examiner, the person qualified to administer flight-checks and oral exams to students prepared for, and seeking, final flight licensing approval.

Jack would spend 40 years, post-military, primarily as a fight instructor, but also taking on pilot examiner duties, as needed. He estimates that, over those years, he has taught over 150 students to fly and prepared them for their licensing (with final tests administered, of course, by another examiner!). Of special significance for Jack is that, having taught his son, Greg, to fly many years earlier, in 2016, one of the final select students that he chose to instruct, and prepare for licensing, was his grandson, 17-year-old Mason Scoggins! Mason, a high school student, residing with his parents in Davidson, NC, spent many summer vacation weeks with his Granddad in Valdosta in order to fulfill the dream of flight held by both. And another crowning achievement came, this one, a year earlier, when veteran teacher Jack Scoggins was named the 2015 Georgia Flight Instructor of the Year, during a formal evening Georgia Aviation Hall of Fame ceremony at the Museum of Aviation, Warner-Robbins, Georgia.

**UPDATE**: Jack Scoggins, at the age of 93, continues to provide aspiring pilots flight instruction with his advanced flight simulator. Also, even more special for us, he is the father of my dear wife, Julie Olsen, DPA .

Lieutenant Colonel Jack Scoggins, USAF (Ret.) a distinguished military veteran, and a life-long genuine American patriot.

# CHAPTER 12

## From Armor's Might to The Call of God Almighty, Chaplain LTC John E. Scott, United States Army

Despite being born at Wright-Patterson Air Force Base in Fairborn, Ohio, and, later, even though he had two older brothers in the Marine Corps, it would take another key life experience before the idea of a military career would become reality for the eventual Army Lieutenant-Colonel John Scott.

He was raised in Xenia, and then Yellow Springs, Ohio, where he would attend, and graduate from, Yellow Springs High School. While there, he became a three-sport athlete, participating in football (quarterback), basketball, and track, excelling in the later (110-meter-high-hurdles). So impressive was he, in fact, that in his junior year, he received a letter from West Point to see if he might like to run college track there with the Corps of Cadets. His mother then arranged for him to follow up by interviewing with the local West Point representative, leaving Scott then thinking about attending the Academy.

In the meantime, he would also receive a track scholarship offer from the University of Cincinnati. Before long, it was decision time. At which point an older, wiser former-Marine brother, reminded him that going to West Point would be like "enlisting in the Army," while running track for the University of Cincinnati would enable him to be "a regular college kid," with Army ROTC there, if that became his interest. Brotherly logic won out. Following high school graduation, Scott enrolled at the University of Cincinnati, where he would run track with distinction and, for the time being at least, put any ROTC thoughts "on the backburner."

At some point in his freshman year, Scott decided to change his major from Electrical Engineering Technology to Criminal Justice. But when he dropped those Engineering classes for a new class schedule, he found himself with only 10 credit hours rather than the full-time 12 hours required to keep his track scholarship! And perhaps unexpectedly, here is where his career path begins to change. Seems that the only two-hour class then still available was 'Introduction to Leadership' within the ROTC program! And it would become a positive turning point. As Scott recalled, he "enjoyed being around his classmates, and enjoyed learning about leadership." So much so, in fact, that he would go on to apply for a two-year Army ROTC scholarship, and received it, as he began his junior year. Scott would continue to run track, while being in ROTC, as well, which "was just like being on another team." As indicated, he really did enjoy his final two years of college, enhanced by his close association with his ROTC classmates, instructors, and the overall military-focused experience.

John Scott would go on to receive his Bachelor of Science degree from the University of Cincinnati in 1997, along with the designation as a Distinguished Military Graduate, with his commissioning on June 14th (Army birthday) as a Second Lieutenant in the Armor Branch.

*Second LT John Scott at his Ranger Graduation.*

His first duty assignment would be at Fort Bragg, North Carolina. But first, it was off to the Armor Basic Course at Fort Knox, Kentucky (November 1997 – March 1998). Followed then, in April 1998, by Airborne School, finally getting to Fort Bragg that June. Once there, all of the
platoon leadership slots were full, so his commander asked him if he wanted to keep the schooling going by heading directly to Ranger School at Fort Benning in Georgia! He accepted and excelled (September – November 1998)! Upon his return, at last, to Fort Bragg, he became second platoon's leader in Alpha Troop, 1-17$^{th}$ Cavalry. Followed in one-year by his move to become the Executive Officer for Alpha Troop, 1-17$^{th}$ Cavalry, a position he would hold for one additional year. Then Executive Officer for Headquarters Troop, moving later to become Troop Commander for two Calvary Squadrons, all while serving in the 82$^{nd}$ Airborne Division at Fort Bragg.

More Army schooling followed. First came the Infantry Captain's Career Course in January 2001(six months), then on to the Combined Armed Service and Staff School (August 2001). Before we get too far ahead, so as not to overlook his previous key accomplishments, following Ranger training, Scott attended the Advanced Airborne School Jumpmaster Course (1998), Air Assault School (1999), and SERE High Risk Level C training (2000).

After his two prior 2001 training sequences, in October of that year, Scott went to the Armor Center at Fort Knox, where he took command of a Cavalry Troop (Echo 5-15 CAV), a position he would hold for the next 19-months. After this training troop command, he was promised a second command, this one to be in a regular Army Deployable Force Command unit. Asked where he wanted to go, without hesitation, Scott selected Third ID at Fort Stewart (Hinesville, GA). Unfortunately, at the time, the appropriate commands there were filled, meaning a three-year wait for a command opportunity. Asked for a second preference of those available, Scott chose Korea!

"So, I ended up going to Korea and got a second command there for Headquarters and Headquarters Company (4-7th CAV), where he would command troops for about 14-months (July 2003-October 2004). Recalled Scott: "I really enjoyed my time and command over there." Considered to be "an unaccompanied hardship tour, I got to pick my next location assignment." So, at that point, he opted to return to Fort Bragg. And here his first service transition would take place, this being from commanding combat arms troops to the Army's "functional area" of Civil Affairs. He would devote the next year to Civil Affairs Officer Training, specifically the Psychological Operations sequence. Followed by the Regional Studies Course, then the Civil Affairs Qualification Course, and finally four-months of French Language School (predictor of world regions he'd likely be assigned), completing that package of studies in January 2006. "I then went immediately to Afghanistan, my very first Army deployment!" remembered Scott. There, although his control branch was still Armor, he was serving then as a Civil Affairs Team Leader (January-June 2006). An assignment he considered to be "really neat." "We did a lot of humanitarian assistance, after talking and coordinating with local leaders, so that was all good."

*Captain John Scott with children in an Afghanistan village.*

All good, that is, with the exception of one very close call. And Scott remembers it very clearly: "There was a mission where the organization I was working for had arrested some people the night before. So, after these arrests, they want to send a civil affairs team back to that same location to assess any damage and reassure the people there that if there's anything the military can do for them, we'd try our best to make it happen. So, we were supposed to go back to this village the next day. And what we were going to do was pass out some humanitarian supplies, do some medical work, and then just talk with the leaders, along with assessing any damage. But in our dining tent that night, our interpreter said, 'hey, you need to know that everybody already knows about the follow-up mission that you're scheduled to go on tomorrow. Even some of the contract local nationals who work here in the dining facility. They know.' So, our interpreter felt certain that our mission was compromised. At that point, I had a decision to make. People weren't supposed to know that we were planning to go in there. I had to let my higher-ups know that our plan was likely compromised. We did a little more assessment of the situation. Finally, I said, hey, I don't think we should go on this mission. Decision made, we didn't go. The next day, some Afghan police did, in fact, go there and they were ambushed on the way into that town. That would've been me and my team! We did end up going back there about a week later. We found out, then, that the insurgents had actually out a bounty out on civil affairs teams and people doing humanitarian work, because they did not want us winning those hearts and minds! Turns out that was the only time we had potentially been in acute danger during our deployment missions."

Meanwhile, it was while he was in Afghanistan that John Scott's second, and most long-term, impactful, Army transition decision began. Which was his serious consideration of now transferring from Combat Arms and Civil Affairs to becoming an Army Chaplain. Here while on deployment, he realized that he was then facing a career crossroads. He had reached the point where he was up for promotion to Major. By Army regulations, had he taken that promotion, the door to his chaplaincy dream would have slammed shut. So, with that key promotion, he could have either gone back to armor or stayed with civil affairs, but chaplaincy would no longer have been an option. While there, in Afghanistan, looking toward his hoped-for future, Scott had actually begun taking his first seminary course online! Along with that move, he also sent an email to get an Army chaplain recruiter.

It was actually way back during collegiate ROTC Summer Advance Camp at Fort Bragg, when he had first begun attending chapel services on Sundays. Scott then shared some of his long-considered thinking about his growing desire to make this key transition. The Army career move he considered now to be the right one for him. Sharing his deeply held reasoning for moving over from combat arms and civil affairs, to full-time chaplaincy, John Scott revealed his deepest heartfelt thoughts: "Back in 1996, I was engaged to be married. My future wife and I shared an infant daughter. I knew then that I just wanted to be a better father. And I felt like having a strong faith in God was the right way to go with that. So, in 1996, I became a Christian. My wife was already a Christian and, you know, I just felt like that kind of faith, that kind of foundation, was good for building a family. And I well-remembered that, back during 1998's Ranger School, I was going to the chapel services that the chaplain there offered in the field, and they just really encouraged me. After one of his services, I actually wrote a letter to my wife and said, hey, I think I want to be a chaplain someday. You know, his services really help, and I think I want to do that for soldiers down the road. So that was in 1998 when I wrote that particular letter indicating my building future desire. And then it was in 2006, in Afghanistan, when I finally got in touch with the chaplain recruiter, began talking with him seriously about it, and then I put in a transfer packet."

The decision made, and approved by the Army Chaplain Corps, John Scott then entered seminary as a hybrid (classes online and in-person) degree program student at Liberty Baptist Theological Seminary, a component of Liberty University, at which point he became designated as a 'chaplain candidate.' For his next three years of seminary study, he would be temporarily removed from active-duty status and assigned to the Virginia National Guard (with drill weekends & summer training). Fortunately, however, his overall

Army service time would continue uninterrupted. It should be noted that, to their credit and commitment, Scott and his wife paid for his three years of seminary schooling themselves. Reason: The Army doesn't pay for seminary, and had he taken available tuition assistance from the Guard, he would have then been obligated to serve a Guard commitment, rather than, as planned, coming immediately back onto Army active duty following his graduation. That simplified the tuition decision!

After his three years of seminary study and graduation, he would finish his 'packet,' and apply back to the Army Board to be considered as an official Army chaplain. So, then, in November 2009, John Scott was returned to active duty as a United States Army Chaplain and was officially re-commissioned as such in June 2010.

You'll recall that Scott's trade-off for pursuing an Army Chaplaincy was giving up, for a time, his pending promotion to Major. Well, that 'for a time,' indeed proved to be quite a lengthy amount of time. As he remembered it: "I stayed a Captain for a very long time. Typically, you wear the rank of Captain for six or seven years. I wore mine from almost going to Major the first time (3 ½ years in the National Guard while in seminary), and then while serving as a Chaplain for about 15-years or so. That's about a little more than twice the normal promotion time!" But he accepted the considerable number of years of delayed promotion because of the positive decision he had made to train for, and then move into, the Chaplaincy.

From September 2010 to May 2011, in his assigned new role as Battalion Chaplain for 2nd Battalion, 325th Airborne Infantry Regiment, 2nd Brigade Combat Team, 82nd Airborne Division, he would do pre-deployment training, followed, then, by an actual deployment to Iraq, in June 2011, as a part of 'Operation New Dawn.' "Our charge," remembered Scott, "was to continue to train, and assist, the local army, along with the local police, so that both were capable of carrying out their own operations. This 'Operation' was intended to be the shutting down of our efforts in Iraq. At one point, there was a chance we were going to be extended past December, but our leaders didn't work out the necessary departure deal, so regardless, in December 2011, they shut it all down and we departed for home. Overall, that deployment wasn't that bad. Although, we did have one of our squads, driving in a vehicle, get hit by an IED. In that event, one of our infantry men was killed instantly and another one was pretty badly injured. That happened right outside Baghdad." Thus ended Scott's first deployment as a chaplain. He would end up spending a total of just under three years (2013) with that unit. In June of that year, he was assigned to be a cemetery chaplain for the Army at Arlington National Cemetery. "While there, I would be conducting anywhere from three to five funeral services, four to five days a week." He would serve at Arlington from 2013 to 2015.

*Major John Scott (second from left) conducted his final funeral service as an Arlington National Cemetery Chaplin, joined by his wife Gwen Scott, escorted after the service by colleagues.*

Then and there, at long last, in March 2015, John Scott received his promotion to Major! Followed in April of that year by a move from the National Cemetery "up the hill to Fort Meyer to become the regimental chaplain for the Old Guard" the special soldier unit (i.e., the Third United States Infantry Regiment), that serves as the ceremonial Honor Guard at Arlington. From 2015 to 2017, Major Scott was the Regimental Chaplain conducting as many as 600 total funerals between those two assignments (Arlington & Fort Meyer), from 2013 to 2017. "It was quite an honor to be up there," recalled Scott. Following those special assignments, he was then selected to attend the Command and General Staff College at Fort Leavenworth, Kansas, for military studies that would last one-year (2017-2018).

*LTC Scott's promotion ceremony (2021), assisted by his wife Gwen Scott.*

After his graduation in June 2018, he went to Fort Bliss, Texas for a two-year assignment as the Deputy Division Chaplain for the Army's First Armored Division. Then, in 2020, he moved over to the Garrison at Fort Bliss to become the Garrison Chaplain there for about six-months. The Army would then move him across the country to the Pentagon in the D.C. area, where he would serve as the Communication Strategy Officer for the Army's Chief of Chaplains. Remembered Scott about that assignment: "It was kind of a neat position. I got to write some speeches for the Chief of Chaplains, along with helping him prepare for those speeches. I was also able to do some social media management, as well as taking some photos and doing videos, then writing stories, just like a regular PAO (Public Affairs Officer) would do!"

*LTC Scott's son Second LT Maliek Scott's graduation from Ranger School (2021)*

Towards the end of his time there, around June 2022, the Army found out that one chaplain wasn't going to be able to take the position for which he had been selected. So, then, a few more chaplains were moved around, with the end result being that they didn't have a division chaplain to come to the Third Infantry

Division. Remembered Scott: "I had already gone through the assessment program and assessed favorably to, in fact, become a division chaplain. But that was supposed to happen in 2023. It was just 2022 now! But the chaplain's branch called me and said, hey, you're going to go a year early, and you're going to go to Fort Stewart!" Thus began his two-year stint (2022 to 2024) as Fort Stewart & Hunter Army Airfield's Division Chaplain, with overall supervision of 35 combined post chaplains, of whom five were direct reports (brigade-level chaplains). As this is written, all too soon it seemed, his assignment here with us would conclude at the end of March 2024.

With regard to his time serving here in Coastal Georgia, Chaplain Scott was asked his opinion of the level of military support he had witnessed over these past two years among the citizens of both Hinesville, GA (Fort Stewart) and Savannah, GA (Hunter Army Airfield). Said he: "I think it's really good. There is some great integration here between the civilian leaders and the leadership in place at Fort Stewart and Hunter. For instance, larger employers in the area are always anxious to interview our people who are retiring, or ending their service commitment, for both employment and internships, as a lot of our folks obviously would like to retire in this area, because, frankly, it is so receptive to both active and former service members. The sense of community we've enjoyed, both on the installation(s) and outside, has been some of the best I've experienced over my two-plus decades of Army service."

*Fort Stewart departure ceremony on March 15, 2024 (LTC Scott on right)*

The conversation then turned to leadership. Chaplain Scott was asked whether he thought strong military leaders were born or made? His response: "I think that, at least to some extent, they're 'made' because there's no substituting for experience. You know, you can be born with a high IQ, and/or athletic ability, and all of those other factors, but it's experience that puts it all into context. So, I do think it's a mix, but I think that the investments that people have made in me have really made the difference. I've been blessed with some fine role models, the really good soldier-leaders that I've had. So, I definitely think that, without their lasting input, I wouldn't be the same person, or leader, that I feel I've become."

Continuing on the leadership theme, the Chaplain was asked if he could think of a male or female senior officer who he remembered as being an outstanding leader, and what qualities made him so? He responded without hesitation: "Yes, there are a bunch, but I will go with my Brigade Commander Colonel at the Old Guard, who's now the Commanding General for the U.S. Army Recruiting Command, Major General Johnny Davis. And, for me, the thing that made him outstanding was that he's just so personable. Whenever he engaged with people, he would do so fully. He never seemed like he was in a hurry. Regardless of the rank of the soldier he was talking with, Colonel Davis made him or her know that he was talking directly to that soldier, even caring to learn about where he or she was from. When he asked: 'How are you doing?' he made eye contact and stopped and listened closely to their response. So, what that did was to make everyone in that brigade, and that regiment, completely motivated to do a good job, and to be loyal to him. And for me, that was just the right model for how to lead a unit. He let his staff handle all of the staff and detail work to be done, so that he could just lead. Being certain that he had the time to actually be the great leader that he was then, and still is."

Then, finally, Chaplain Scott was asked what best advice he would give to a young, enlisted soldier, or officer, just beginning what he or she hoped would become a military career. Said he: "The advice would be the same as that I give to a lot of the chaplains just coming in by a direct commission, that is, those without basic training, those with no prior service, and perhaps my biggest advice would be to just ask questions. There is so much knowledge and experience held by those serving all around you. For instance, in my case specifically, to those I supervise, my advice is to take a close look at the issue or problem in question, and if you can't figure it out in 5 or 10 minutes, just ask me!

Shoot me a message, give me a call, or just come by the office. Let's work it out together. But don't struggle with the issue or problem for a long time. It's perfectly fine to simply ask questions."

Wondering if there was anything else that Third Division Chaplain Scott might like to share, he thought for a second and thoughtfully added the following to our discussion: "I guess the one thing further is that I really have enjoyed the chaplaincy. It is a unique position. People naturally think of the religious aspect of it, which is absolutely very important, but the day-to-day thing that I feel we bring to the Army is the human connection. Commanders and other leaders always have a lot of different things on their minds, meaning that they can't always be as involved in the human dimension as they'd like. And that's where the Chaplains and our Religious Affairs Specialists (the enlisted side of the chaplaincy branch) come in. To be there for people in times of struggle or perhaps even in some dark times. We're there to listen and give advice. So, we in the chaplaincy have to know about the entire Army enterprise in order to be able to assist people to get linked up with the help they may need. And that has been the most rewarding for me, assisting soldiers, their families, and our Army civilians, with issues and problems occurring in the human dimension."

Concluding our discussion time together, here is some additional information for the reader to know about this very impressive Senior Army Chaplain. He has earned the following graduate certification and degrees: Certificate in Communications and Business Administration, Liberty University (2021); Doctor of Ministry, Wesley Theological Seminary (2020); Master of Military Arts and Sciences, U.S. Army Command and General Staff College (2018); Master of Divinity, Liberty Baptist Theological Seminary (2014); Master of Religious Education (2009) and Master of Arts Religion, (2008) Liberty Baptist Theological Seminary.

LTC Scott's military awards and decorations include: the Meritorious Service Medal (6 Oak Leaf Clusters), the Army Commendation Medal (3 Oak Leaf Clusters), the Army Achievement Medal (2 Oak Leaf Clusters), the Afghanistan Campaign Medal, the Iraq Campaign Medal, the Global War on Terrorism Service Medal, the Senior Parachutist Badge, the Air Assault Badge, and the Ranger Tab.

Chaplain John Scott married his wife, Gwen, in 1996, and they have been blessed with four children, daughters Jonni, Kara, and Sharon, and son Maliek. Chaplain Scott is presently approaching 27 years in Army service to America.

**UPDATE:** This fine career Chaplain has now moved on to a senior assignment in the Washington, D.C. area.

# CHAPTER 13

## From a Cold War to a Hot One
## Lieutenant General E.G. "Buck" Shuler, Jr.,
## United States Air Force  (Ret)

"Look out for that helicopter!!!" yelled his back-seater.  Up until that jolting instant, their mission had been a challenging, but ultimately, satisfying one.  It had come early in U.S. Air Force Captain E.G. "Buck" Shuler, Jr.'s tour flying F-4's in Vietnam.  "Operation Niagara" was our intense air-effort to end the North Vietnamese Army's all-out assault on the surrounded U.S. Marine base at Khe Sanh (South Vietnam).   On that day, Captain Shuler was flying as wingman to Air Force Major Fred Caldwell.  Caldwell's F-4 was loaded with 750-lb. 'high-drags' (non-precision 'dumb' bombs, this version designed for close-to-ground release). Shuler's with "wall-to-wall" napalm (8 'cans').   A Forward Air Controller (FAC), flying observation over the combat area, radioed word of a reported active enemy gun position either on, or just below, a ridge line running above a deep river canyon, located close to Khe Sanh.

For this attack, the FAC (call sign: 'Rash 0-3') noted that Caldwell's bombs wouldn't be the choice, but Shuler's napalm would.  "So he (FAC) marked the spot, and I rolled in, running south to north" said Shuler.  The preferred altitude/speed combination for delivering napalm, at the time, he noted, was at about 450 knots from around 450-feet.  Shuler made his target-run and dropped napalm on the specified site. "But it blew over the position, rolling over the top of the gun emplacement and down the ridgeline, from what the FAC described," he recalled.  The FAC then told him to try it again from the reverse direction.

So, Shuler flew back around the target area, coming in this time from north to south. But in order to do that, he had to fly down low through that river canyon, a heavily defended one at that. "I was right down close to the water, then rose back up and 'splashed' the napalm just below the ridgeline," he recalled. That did it. The FAC radioed to Shuler that the gun position was now 'silent' (100% effective!). As Shuler's primary concentration was then fully on his exit path, pulling off on a right-climbing turn, increasing his speed as he did to about 500-kts., that was the instant back-seater, Captain Stan Czech yelled to warn about a helicopter directly in their path. "I immediately pulled-wide to avoid colliding with a damn Chinook, that had a 75-mm howitzer hanging beneath it. I flew past those guys, just scalding them, and that had to have rattled their chains!" remembered Shuler.

With that certain multiple-fatality, near-death experience still vividly in mind, Shuler stated that "the most dangerous thing about flying in South Vietnam, was not ground fire, although small-arms were certainly down there, but really no big stuff, except up around Khe Sanh. The biggest danger was mid-air collisions. Because no matter where you were, you'd put in a piece of ordnance and, immediately, every damn airborne helicopter and other aircraft wanted to come see and watch the show! So, it became a see and be-seen, and try to avoid, type of thing. We had a number of mid-air collisions because of that," recalled Shuler. He also remembered that another thing he and his fellow fighter/attack pilots had to watch out for were B-52 strikes, even though they were generally pre-briefed on those big-bomber missions.

Captain Shuler's repeated efforts to eliminate that ridge-top enemy machine gun position, flying fast and very low through small-arms fire in that enemy-infested deep valley, and achieve the goal, earned him the Distinguished Flying Cross (April 5, 1968).

Moving the narrative well back, now, to the point where his distinguished Air Force career began, then-Second Lieutenant E.G. "Buck" Shuler, Jr., commissioned at The Citadel (Charleston, SC) in 1959, successfully progressed through the Air Force flight-school sequence, pinning-on his Silver Wings on September 2, 1960. As with most in his graduating class, he was then assigned to multi-engine training. One of only four new pilots from his class selected to fly B-52's, Shuler's clear preference, since he definitely wanted an aircraft "that was operational," that is, actively involved in front-line combat missions.

*Cadet Buck Shuler, Senior Year, The Citadel*

His route to the B-52 flight deck took him first to ground school, followed by additional flight training, then nuclear weapons school, and, finally, survival school. Note the nuclear weapons training component. With the Cold War then well underway, America's B-52's formed the third leg of our deterrent nuclear triad (air/submarine/ ICBM). With so many decades between then and now, even for those of us who lived through those times, it's hard to remember that, for about four decades, how unrelenting, and often times tense, was the stand-off between America and the Soviet Union, within the framework of mutual, nation-ending destruction, the latter, nobody-wins scenario, being the only thing that kept that Cold War from turning devastatingly hot.

Shuler and his five-man crew (aircraft commander/pilot/radar-navigator/navigator/electronic warfare officer/gunner) were paired prior to their arrival at the Strategic Air Command's (SAC) Carswell AFB in the Spring of 1961, and were assigned to the 9th Bomb Squadron, 7th Bomb Wing. Shuler would serve with this same "Select" S-22 crew for the next two and a half years (Major Melvin Apel/Aircraft Commander; First Lieutenant E.G. Shuler/Pilot &

Deputy Aircraft Commander; Major Al Herman/Radar-Navigator; Captain Ruel Branham/Navigator; Captain Guy Kreiser/Electronic Warfare Officer; and Master Sergeant Lonnie Plummer/ Gunner).

*B-52F,*
*Select Crew S-22 at*
*Carswell AFB, Texas,*
*1st Lt. Shuler is the pilot and*
*Maj. Melvin Apel is the aircraft*
*commander.*

*(Shuler, back row,*
*second from the left;*
*Apel, back row,*
*third from the left).*

Following flight check-out requirements, now-First Lieutenant Shuler and his crew began pulling Cold War-era ground-alert and airborne duty. "Of the 16 B-52's in the Wing," he recalled, "half were on ground alert at all times," a continuing requirement imposed by the on-going tension. Which would soon intensify even further, as international concerns and tensions rose due to those unacceptable missile shipments from the Soviets to their Caribbean surrogate: Cuba.

The 1962 Cuban Missile Crisis placed additional "alert" stresses, not only on the SAC bomber crews (B-52's and B-47's), but on their families as well. "Family life in SAC was tough, demanding a special kind of spouse," he said. An assessment certain to be shared by the thousands of military wives, husbands, and families who've experienced, and had to adjust to, back then, as now, the lengthy (or shorter but more frequent) deployments demanded of our skilled and courageous warfighters, and certainly so from the aftermath of 9/11 forward.

Shuler clearly remembers that when President John F. Kennedy appeared on national television, on a Sunday afternoon, to speak about the impending defense crisis in that all-too-close-to-us island nation, Shuler was then sitting on, by-then, routine ground-alert at Carswell AFB (seven-day cycles, pulled typically twice per month, to include ground and airborne, as well as training flights). "When this thing really looked serious, as the alert-status was ratcheted-up two-days after the President's speech, I went home, packed a clean flight suit and more clothes, then went back out and began additional ground alert," he recalled.

At that point (1962), Strategic Air Command immediately "cocked" its entire bomber fleet and terminated all training. As Shuler explained: "Every B-52 and B-47 was put on the ground and loaded with nuclear weapons. Some of the 47's, (having shorter range than the 52's), went into a cautionary defensive dispersal plan, flown to pre-set overseas airports, as well as domestic, to just sit there under nuclear alert." SAC then instituted a $1/8^{th}$ airborne alert concept. Meaning for those at Carswell, two B-52's would, henceforth, be in the air, seven days straight, around the clock.

And those weren't leisurely domestic flights. Far from it. "In those days," said Shuler, "we were flying a route that took us from Fort Worth, towards the Mediterranean, then refueling mid-air as we hit the northwest corner of Spain, flying on in by the island of Majorca (off the eastern Spanish coast), before coming back out by Gibraltar, for a second mid-air refueling, bringing the aircraft up to full tanks, then all the way back to Fort Worth."

Stretching around the Soviet Union was an imaginary line called the 'H-Hour Control Line.' All of our bombers would be required to refuel prior to reaching that invisible point, for obvious reasons. In addition, the perimeter would provide a timing and positioning check-point, a last line of control for command authority back in the States, for assurances that all was in order with the armada of American aircraft, as they prepared to penetrate actual Soviet airspace and deliver their nuclear weaponry.

Those 'what-if' preparatory missions generally lasted about 24 hours. The longest one he flew went about 25-hours and 20-minutes, as he recalled. Shuler and his colleagues at Carswell remained in that same flight rotation for the next thirty-days. During that specific period of international tension created by the threat of Soviet missiles in Cuba, Shuler flew six of those rigorous round-the-clock airborne missions, remaining on ground-alert status in between.

And talk about intense. America's Air Force "had four B-52's refueling near Spain every 30-minutes, with pre-assigned targets ready to go (in the Soviet Union, Warsaw Pact nations, and others). All that was needed to initiate our nuclear response was the 'Go Code,' which could only be transmitted by the U.S. President," said Shuler. The SAC fleet eventually came out of that particular period of military intensity, with its unthinkable potential for global destruction, when, thankfully for those of us alive, back then and today, the Soviets 'blinked.'

At this point, it's important to note that the Commander in Chief of Strategic Air Command had the authority to launch the bomber force for reasons of survivability, but only the President had the authority and responsibility to execute the Strategic Integrated Operational Plan committing the forces to the use of nuclear weapons.

While we were normally in a Cold War Defense Condition 4, during those years, the Cuban emergency had elevated the 'DefCon' all the way to Condition 2. "It was a dicey period of time, as I recollect. I thought we probably had a 50-50 chance of going to war. As it turned out, we didn't," said Shuler. And thankfully so, for then-current and future generations of Americans, as well as all peoples who would have inevitably suffered the wide-spread and lasting brutality of nuclear war, certainly something we've been conscience of here in the U.S., watching international frictions develop, along with justifiable concern for nuclear weapons in the hands of rogue states or terrorists.

One of SAC's other taskings during that time of international tension, was the use of its airborne-alert aircraft, when crossing the Atlantic, to be on the look-out for Soviet ships, helping our Navy spot and track any shipments of munitions, missiles, and light-bombers to Cuba.

As the intensity of the crisis in Cuba thankfully became less so, with Shuler and his same crew from Carswell AFB then moving on to Dyess AFB (Texas) in September, 1963, the alert flights continued, but were reduced back to one B-52 at a time, and with a different flying route, this one heading from Texas to Massachusetts, refueling, then on up to Greenland (making communication checks with the Distant Early Warning (DEW Line) personnel, proceeding close to the North Pole, before turning back south to Point Barrow, Alaska, refueling, then out to the end of the Aleutian Islands, before turning back down the U.S. West coast to home station at Dyess AFB. As he recalled, with an understatement: "Long missions! With my crew, we set up a routine where we'd keep one pilot in the seat, with either the gunner or the EW Officer monitoring the radios, leaving one of the navigators in the seat, as well. Then the rest of us would try to get four hours of rest. You couldn't actually sleep on the aircraft due to the engine noise (and on-air mattresses, no bunks!). You could only rest, that was about it. Then we'd swap-out every four-hours. The only exception to the rest rotation was that all crew members had to be in their seats during the critical phases of the flight: take-off, mid-air refueling, and landing."

As Shuler remembered it, the 'fly-away' cost of each B-52 produced when it first became the primary bomber for SAC (1955), with its crew of six, was about $ 11 million each. "That was all a pretty heady responsibility for a 27-year-old Air Force Captain, and, by then, an aircraft commander as well, at a time when the average age of the squadron's aircraft commanders at Dyess AFB was 39.5-years!! Real old heads with a lot of experience," said Shuler. Not only had he earned the aircraft commander designation at a remarkably early age and rank, but he also earned the assignment of commanding a "lead" crew, meaning one at the very top of assessment and mission responsibility. The crew progression sequence back then began with 'Ready,' advanced with qualification achievement to 'Lead,' and then finally to 'Select,' based on performance proficiency. As Shuler indicated, pretty 'heady' stuff, indeed!

He recalled that the basic weight of the B-52 was about 184,000 pounds. The actual weight at take-off, with full combat load (fuel, crew and munitions), rose to about 456,000-pounds. Like the airfield at Hunter Army Airfield in Savannah, Georgia, back then, as it remains today, SAC runways were normally at least 10,000-feet in length, with an additional 500-feet of 'overrun" at both ends. "Even with normal training weight," recalled Shuler, "many take-offs had to use water-injection in each of the eight engines to get airborne. I can remember having to use a full 10,000 feet on very hot days at Dyess. Even with helmets on, you could clearly hear the engines, and as you rolled down the runway, the noise from the straining engines increased to a high-pitched scream."

When asked about the degree of difficulty in piloting such an enormous aircraft, in both weight and exterior size, he responded: "The analogy that many crew members used to describe it was that it felt like you were driving an eighteen-wheeler. Big, and with no boosted controls, it was very tiring to fly." And beyond normal flight demands, with the notorious Texas crosswinds, landing a B-52 safely back at base could be made additionally stressful.

Fortunately, the aircraft did have the ability to deal with that condition, to a point! "The B-52 had crosswind 'crab-landing' capability. The system moved the quadri-cycle gear-trucks up to a 12-degree maximum, based on the wind readings from the tower. I can remember landing a B-52 looking out the pilot's side window, which is rather spooky! But that capability did make a landing there possible, rather than having to divert to another SAC base," he said. "For all of those reasons, I found the aircraft to be a real challenge, not only to fly, but also to lead and coordinate the crew, so that, if it ever came to that eventuality, we got those bombs on-target. But one of the greatest stresses with many of those B-52 missions, especially the Cuban-era, high-alert overseas ones, was the sheer fatigue from their duration, often 24-hours+, adding enormously to the normal, everyday, physical effort it took to fly that aircraft," remembered Shuler.

Going back to the air-attack munitions load, during the Cuban crisis, it's important to realize that Shuler's aircraft, along with many of his sister '52's, routinely carried two nuclear weapons on their prescribed overseas missions.

In addition, some of the bombers actually carried the 'Hound Dog' cruise missile on their flights to the Soviet Union or related potential targets, if it became necessary to fight their way into the target. The missiles provided the muscle for the Soviet bloc attack plan, which was to "shoot and penetrate," recalled Shuler.

Along with the airborne bombs and cruise missiles, the other potent element for our nuclear response would be the Intercontinental Ballistic Missile. Depending on ground or sea (submarine) launch point, the ICBM flight time to target would be about 30-minutes, leaving little time to avoid mutual destruction between the two powerful nuclear nations.

Hard to imagine now, but ultimate reality for crew, nation, and humankind back then, in a so-called Cold War, it was always just one action by the President away from turning irreversibly hot! As such, it goes without saying, but regardless bears repeating, that it was a time of 24/7 tension and fatigue for crews tasked with making those SAC flights, armed with the capability to produce incredible devastation. And due to the classified nature of these potentially deadly defensive flights, especially during the actual period of the Cuban Crisis, the American public was largely shielded from the stark reality that their nation and planet were wavering, as never before, on the very edge of its continued existence. But there was certainly awareness, though with widely varying levels of concern, about what civilians here could do, unrealistic as it may be.

Actually, well before the Cuban missile situation, during the 1950's especially, in the event of that feared nuclear attack, school students were drilled in "safety" procedures, such as lining the school hallways, or simply putting heads under desks (clearly desk construction and survivability then were superior to today's models!). All ineffective, of course, but making at least a naive effort to plan something, anything, for the nation's children before the threat became reality.

And, for America's concerned adults (as all were who were paying attention), month after month, magazines like "Popular Mechanics," "Popular Science," and doubtless others, featured plans for building a family's own, in-ground, backyard fall-out shelter. Some were basic, little more than a large hole in the ground, with a reinforced roof and filtered air-intake), stocked with some non-perishable food, and cots for sleeping (bathroom solution primitive). Others, for the more prosperous, were well-outfitted and comparatively lavish, ready for the presumed longer haul. Then, the moral/ethical dilemma of the day for shelter owners became, what to do if someone knocked frantically on your shelter door once the bombs had hit, as the forewarned deadly radiation spread. Would you endanger your own family by opening the door to try to save others? The universal answer became, pretty much, NO!

All of this intensity of concern and planning peaked in the 1950's and 1960's, overshadowed later on in that latter decade, by the public's increasing desire to get American troops out of Vietnam, along with other political/social/economic issues closer to home. But the Cold War would continue, though perhaps with lessened intensity and public attention, for many more years, with the SAC ground-alert status actually continuing until President George H.W. Bush finally ordered the stand-down in late 1991. And one historic note for those in Coastal Georgia: The test program for SAC's ground-alert status, employing the nuclear-capable B-47 aircraft, actually began in 1958, at Hunter Field in Savannah, Georgia.

Returning to those extreme crisis days, brought to peak intensity by the Cuban Crisis, the around-the-clock U.S. air missions of our lethally loaded aircraft would take off, each with two targets assigned, in the event the alert-mission turned active. Depending on real-time circumstances, target change-orders might well come while in flight. One primary bomb target assigned to his crew remains clearly fixed in Shuler's mind, and for obvious reasons: "I can remember having the southwest corner of the Kremlin as our aiming point." Recall that those (nuclear) bombs on the B-52's existed well before the precision guided ones of today. If the target was a specific military facility, a pilot would actually have to fly to, and drop the bomb, there! Important to note, also, that our bombers would be flying to, and hopefully back from, the target area <u>without</u> American fighter escort. He recalled that, if they survived the flight through Soviet airspace, the bomber crews were assigned headings for post-strike Allied bases overseas. "Once you delivered your weapons, you were kinda on your own," said Shuler.

During those extended years of ground and air alert status, all of our military nuclear capabilities were grouped under a Single Integrated Operational Plan, known by the acronym 'SIOP.' It "coordinated <u>all</u> nuclear weapon strikes, B-47's, B-52's, ICBM's, and SLBM's (submarine launched), as well as planned the time on-target, how many weapons to be aimed at each, and de-conflicted the routes of the bombers, so as to avoid previous bomb-strike areas."

"Basically, the first priority," remembered Shuler, "was to neutralize the strategic rocket forces and bombers of the Soviet Union. Priority #2 was to take-out their leadership. And the third target was the Soviet's conventional forces, the Russian Army, and the Warsaw Pact armies, assuming that they would then be attempting to penetrate Western Europe." Again, all pretty scary to contemplate, with sincere gratitude that none of that, in the end, had to occur. May the same hold true for the political and military challenges we face today!

"We watched the Soviets 24/7," said Shuler. The daily intelligence briefings always focused heavily on the Soviets, and questions about what their forces were doing: "Were they in garrison, were they exercising, were they deployed? Were their strategic air forces at home bases or had they deployed up to their Arctic Circle bases, which would indicate a possible impending strike," recounted Shuler.

"When we weren't on ground-alert, we were target-studying, completing training requirements, and planning training missions. By the way, our training missions in those days were anywhere from 8 to 14 hours long."

"With the training missions, we'd do essentially the same thing we'd do in an actual war mission: Take off, then, later, do an in-flight refueling. When flying in the dark, we'd perform a celestial-navigation leg. And we'd always do low-level penetration, which in those days, and with those non-satellite-guided weapons, low-level was required for accurate delivery," he recalled vividly. There were sophisticated (for the time) practice target-scoring sites, located throughout the United States, which could provide the aircraft real-time, accuracy-on-target feedback.

"A SAC crew member work week typically ran as many as 110-hours. And you were constantly being tested and evaluated. Written exams, and then on alert status, we'd have to take a positive control test to demonstrate that we fully understood what we were expected to do if we got the 'Go-Code,'" Shuler recalled. So then, in flight, how would that activate notice be received?

"The Go-Code was transmitted in a message format, with alpha-numeric numbers and letters. The code was embedded within that sequence. Recall that on a B-52, there were three primary crew members: aircraft commander, pilot, and radar navigator. Each had the actual code sealed in a small, plastic envelope, worn with their dog tags around the neck. Each would copy down the just-transmitted alpha-numeric code, then break-open their individual envelope and compare it with what was in the 'Go-Code' message. All three had to be in agreement. If so, then it was a valid, authenticated message, and you were to carry out your instructions," he remembered.

Later on, with the need for imposing even more security, given the obvious gravity of the situation and the catastrophic result of a mis-step, the authentication information was contained within a metal box sealed by three locks. With the 'Go-Code' received, each of the three primary crew members, with their own individual lock, would go separately to the box and open their lock. Again, all three had to agree that the code in the box matched the code received in-flight, before activating their pre-set instructions, meaning to actually proceed to their designated targets and release their nuclear bomb load.

Generally, radio silence was observed on the overseas, what-if, alert missions, with the exception of pre-determined times for brief, in-flight check-ins with SAC, via high-frequency radio, to confirm that the aircraft was 'operation-normal.' Years later, direct satellite communication would replace reliance on high-frequency radio use. Was there ever a time, beyond all of the preparation and practice, when Shuler thought they might actually receive a real mission alert? "Yes. We had two events that really got our attention," he recalled.

To set the stage for those 'events', there were three types of message formats that could be received by Shuler and his fellow aircraft commanders. Color-coded 'Dot" messages that indicated either training information, or notification of a change in SAC alert status (for instance, the order to taxi to the head of the runway, then leave engines running or shut down, depending on perceived Soviet status), or, ultimately, a final 'Dot' which was an actual combat-triggering message, giving the aircraft the 'Go-Code' for either the send-off to war, or instruction to fly east to an initial control point for further instruction, that location varying with specific combat targeting.

Should a real-world, final 'Dot' message/'Go-Code' be received, while underway in flight, and now well beyond the initial control point, with a mandatory mid-air refueling completed, the next pre-established command and control clearance was called the 'H-Hour Control Line,' an invisible ring stretching around the total circumference of the Soviet Union. This imaginary perimeter line in the air was the final critical check-point used by the 'SIOP' (Single Integrated Operational Plan) to insure that all of our bombers heading to war were still on-schedule and that their flight positioning remained de-conflicted (i.e., adequate flight path separation within the American armada) according to that overall attack plan, as the fateful final run into the set target(s) began. "You had to cross that final, critical control line (on time), so as to strike your target at the precise time that SIOP indicated," said Shuler.

Returning to those events that remain vivid in his mind. "For that first one, we were on ground-alert (at a time not during the Cuban Missile Crisis, but still very much in the Cold War). In the dead of night, the claxon goes off. We jump up, get our flight suits and boots on, get in the vehicle, and head quickly to the flight line. As the aircraft was powered up, we got on the radio to hear the senior controller at the command post reading an alert-status message, instructing response aircraft to head to the runway and prepare for take-of. Our radar-navigator kept saying 'it's just a practice.' I had to quickly remind him that this was NOT a practice message! This is real world! And so, we did exactly what we were instructed to do, we taxied on out there," recalled Shuler. Fortunately, although a very real alert, for Shuler and crew, this one proved to be a false alarm.

He later learned that some geese had "spooked" the Ballistic Early Warning System (Canada/Greenland), which was interpreted by NORAD to be a possible penetration by Soviet aircraft. Thus, the real world/real threat reaction was triggered, until it was determined that NORAD had, thankfully, only been falsely alerted by Mother Goose!

But Shuler well-remembered that it was, back then, "so early in the ground-alert program, that when the senior controller radioed that we could all taxi back to our assigned parking area, nobody moved. SAC had not yet provided for an authenticated message system to get us back to the alert area and to stand-down!" Illustrating the power of training and strict adherence to procedures, despite what proved to be the benign nature of that particular incident.

"The next thing that happened was when I was a Wing Commander of the 19th Bomb Wing at Robbins Air Force Base in Georgia (1979-80). Again, in the dead of night, I sat straight up in bed, because I heard the alert force bombers and tankers start engines! I knew immediately something was wrong, because, with any exercise, the Wing Commander is always alerted, so he can be out there to supervise. Before I could even get up, the red phone which sat at the side of my bed rang. My senior controller said, 'Colonel, you need to come to the command post ASAP.' I got into my flight suit and jumped into my vehicle. As I drove out, I went past the alert area. I could see the engines running and the lights flashing. As I arrived at the command post, it looked like a real-world situation," recalled Shuler.

Fortunately, yet again, a false alarm. At NORAD headquarters, a pre-programmed practice tape was inadvertently played, and it went out to the entire SAC network.  Merely an exercise tape, not a real-world situation after all.  But true to their repeated training and practice, "it caused a normal reaction, a reaction just as it should have," said Shuler.

Regarding practice ground-alert exercises, Shuler reviewed the two types then in use for the SAC bomber crews.  "You had a start engines, but stay in position.  That was called a 'Bravo.' You had a start engines and taxi.  That's a 'Cocoa.' With that one, you go to the 'hammer-head' (final stop prior to entering the active runway), get clearance, pull onto the runway, set take-off power, roll 400 or 500 hundred feet, then, in a normal practice scenario, as directed, cut back power and return to your normal parking spot."

Looking back at those continuing and incredibly tense Cold War years, one simply cannot comprehend the understandable stress experienced by those B-47 and B-52 crew members (along with those others manning the remaining two-legs of our nation's nuclear triad) should a 'Go-Code' activation and authentication have ever actually been sent for real.  Yes, they were all well-trained, repeatedly tested (including the maintainers), and thoroughly screened for these missions, and would have, without hesitation, carried them out as ordered.  But the reality of what faced them (along with the likely obliteration of all that was below them), and, conversely, the obvious concern for what America might look like, after the Soviet pre- or counter-launch. And specific concern, as well, for the status of their families (alive or dead), when, and if, those air crews were actually able to return from what could quite easily have become a one-way mission.

All of that, no longer fictional "War-of-the Worlds" (Orson Welles' 1938 radio drama), but very real scenario, decades back, is hard to imagine, just as, regrettably, it is today with the increasingly unstable, aggressive, nuclear-reality that lurks out there before us, and our allies, yet again.  Back then, in those turbulent, dangerous-in-an-instant, early-1960's, thank heavens those authenticated 'Go-Codes' were never received by courageous air crews flying our nuclear-armed bombers.  And may there never be the necessity of our current B-52, B-1 or B-2 pilots, or other delivery vehicles, receiving such today, and frankly, ever after. Nuclear weaponry brings annihilation of structures, and all things living, within its kill-radius, abruptly shortening eternity's timeframe in, and to, an instant.

Before Captain Shuler was transitioned by the Air Force from multi-engine bomber duty to piloting jet attack aircraft, some parting facts and experiential thoughts on the B-52. As it happened, its first recorded flight took place, corresponding by chance to its numeric designation, in April of 19<u>52</u>! The B-52's went operational with SAC in June, 1955. That powerful aircraft was created <u>specifically</u> to carry nuclear weaponry during the Cold War and was capable of transporting up to 70,000-lbs of ordnance.

The B-52 was designed as a high-altitude bomber. True to that intent, "I can recall making simulated bomb runs at 48,000-feet," said Shuler. "But in the late 1950's, when the Soviets dramatically increased and improved their air-defense capabilities, including surface-to-air missiles, and early-warning radars out the 'gazoo,' all around their major cities, like Moscow and Leningrad. We then began flying the B-52 in a low-level regimen on practice missions, developing our potential for sneaking under those radars. Doing so, however, put a lot of wear and tear on the aircraft, accelerating the need for, not just routine, but major maintenance." Regardless, lacking any fighter escort, doing all in their power to make it to the target as undetected as possible, for as long as possible, made the low-level flight approach and delivery, mandatory.

Cold War nuclear devices were, by then, more compact in size than our Pacific War originals, but still capable of delivering a devastating punch. Rather than the detonate-on-impact World War II version, upon delivery, the weapons then were 'drogue-parachute-retarded.' "You would release the bomb(s), and a timer would kick in," explained Shuler. "The bomb(s) would float down and land in the target area, sit there and tick until the timer ran down, at which point the weapon would detonate. In the meantime, we would have the time needed to get away."

Even though a nuclear bomb would be delivered on-target with a delaying-device, rather than World War II's detonate-on-impact version, what, then, was the specific SAC plan for getting aircraft and crew clear of the explosion site? Shuler well-remembered the prescribed post-bomb-delivery procedure: "Once the weapon(s) were delivered, you were on your own. You were working for yourself! We would immediately make, what was called, a 'break-away turn,' left or right, roll the aircraft into about a 60-degee bank, pull it around to about 180-degrees, then fly away from the detonation point, at full-throttles to escape the frag-pattern."

"When Tibbets delivered the A-Bomb on Hiroshima, he, too, did a break-away turn," said Shuler. "But you still felt the shock-wave. It's gonna get you, one way or another. The break-away procedure was designed to diminish it to the point where it wouldn't knock you out of the air. While Tibbets dropped his bomb at a lower altitude than ours would've been, it was actually a <u>much</u> smaller impact weapon than the ones we were scheduled to deploy."

"Early on, even though the weapons were drogue-retarded," said Shuler, "we had to deliver them using two different tactics. One was called 'short-look;' the other a 'long-look.' The short version, depending on the weapon, meant that you would go in at about 500-feet, and then, just after the 'IP' (Initial Point, the formal start of the bomb run, a designation dating back to WW II), start a climb to 8-to-10,000-feet, deliver the weapon on target, then make the turn described earlier, while descending back down to the 500-foot regime. With some bombs, the larger ones, you'd do a long-look, beginning with the same procedure, hitting the IP at around 500-feet, then climb to 18-to-20,000-feet, deliver the weapon, and immediately turn, descend, full-power, again hopefully escaping the frag-envelope of the weapon. Later, the bombs got even better, where you could just over-fly the target and drop the time-release weapon, then fly away," recalled Shuler.

To further assist the safe fly-away of our B-52's and B-47's, special exterior paint was used. "The undersides were coated with an anti-nuclear radiation paint. The resulting white underbelly was applied to reflect the nuclear energy released," he said.

After the rapid, hopefully safe, departure from the detonation area, each mission had a predetermined, post-strike recovery base, generally somewhere in Southwest Asia or other Pacific Rim locations. If the mission plan called for penetrating the Soviet Union on a more northerly route, recovery would then most likely be directed more toward one of the Scandinavian nations.

"If you made it to a pre-planned base, but found it destroyed," said Shuler, "you'd then search for an alternate in the region, either military or civilian. Land and scrounge whatever fuel you could get, beg, borrow, steal, whatever it took. Then, if your airplane was still operational, you were scheduled to fly back to the Continental United States." However, if there was no ability to re-fuel during the return flight, they would not be able to get back to the U.S., in which case they were to "hunker down" in a designated, so-called "safe area" to await the end of the war. After which, whenever possible, remaining U.S. forces would be dispatched to try to locate and rescue our aircrews on the ground. There were identifiers to be used for validation, followed by likely evacuation by helicopter. Remembering, once again, all of those post-strike what-if's and fend-for-yourself contingencies, Shuler then commented, with wry understatement: "Those were not very comforting thoughts!"

But given that scenario, for whatever reason, not being able to make it back to America, what about capture? Shuler's response: "With post-strike damage so horrific over there, frankly, we didn't worry about being captured." The harsh reality was that likely many of the bomber crews would not have survived these missions, or in any event, would not make it home to what remained of our nation anytime soon.

On a more positive note, assuming a return to the U.S. was feasible, Shuler recalled that "we had a very detailed re-entry program. Now, understanding nuclear warfare, and this being our first response to a Soviet strike, we knew that, by then, most, if not all, of the 40+ SAC bases, plus Washington, D.C. and SAC headquarters, would've been destroyed in a nuclear exchange. So, then we had a list of options, depending on our routing coming back into the U.S. For instance, if we were coming back across the mid-Atlantic, options would include

South Carolina or Georgia. You'd look for a place like Atlanta (Hartsfield-Jackson), and land there.

We'd have had recovery teams dispatched to those major, alternate locations at the start of the conflict, with maintenance guys, nuclear weapons for re-arming/re-loading, etc. Then you would await your assignment to fly another strike mission, or whatever they wanted you to do," he remembered.

Assuming, again, that SAC headquarters would've been destroyed at some point during the first flurry of Soviet missiles, any follow-on flight or return strike orders would have originated from the sky. Flying around the clock, throughout the Cold War years, was a command-and-control aircraft, the electronics-laden EC-135 (code name: 'Looking Glass'). With an Air Force Flag Officer always on board (Airborne Emergency Action Officer), 'Looking Glass' became the command authority for returning bomber subsequent orders, whether to find a functioning airfield, land and stay, or to land, take on fuel, get food, re-arm, and head back for a second strike overseas. In addition to coordinating the whereabouts, welfare, and any follow-on missions for our bombers, in-silo ICBM's could actually be launched from 'Looking Glass' as well, if normal ground controls were incapacitated.

And that need for repeat strike missions was all dependent on how important the specific targets were, and how effective our first ICBM and aircraft had been putting nuclear weapons on target, that is, each having successfully delivered their assigned 'PK' (Probability of Kill). The plan, then, was for the operational bombers to return to the continental U.S., if possible. Easy enough to say, type, and read, now over 50-years after the fact and well away from the fray. But frightening beyond comprehension to think about the post-nuclear reality that might have been, given that the Soviet goal quite likely would have been the destruction of our nation, as well. Made all the more intense by the fact that "our stated nuclear philosophy, at the time, was that we would take and absorb the first blow (by the Soviets), and only then, respond. That's why we sat nuclear alert seven-days at a time, and why we practiced these airborne alert missions (aka 'unexecuted combat missions') repeatedly and so intently," remembered Shuler.

With 650-700 B-52's then in our arsenal, 1/8th of them (80+ aircraft) were always in the air during the Cuban Crisis, flying their pre-assigned airborne-alert training missions, rather than about half that many (40+) flying at all times during the more 'normal' Cold War intensity.  Due to their more limited range, the nuke-capable B-47's were mostly prepositioned overseas (North Africa, United Kingdom, Spain, etc.), reducing their required strike distance, while sitting alert at rotating 90-day intervals.  Also at that time, some of the B-47's were loaded, not with bombs, but with electronic surveillance equipment, as they flew continual reconnaissance missions around the Soviet perimeter.  As you might imagine, these flights sometimes did attract the attention of Soviet fighter jets. With only a 50-cal. tail gun for defense, the Soviets did manage to shoot down one of our B-47's, resulting in some of the crew members being killed, while others were imprisoned.  That particular event was chronicled in the book, Little Toy Dog (William L. White).

Along with surveillance from the pre-positioned B-47's, the U.S. also had RC-135's doing reconnaissance flights up and down the Soviet coastline and borders.  These specialty aircraft were equipped for electronic surveillance, pinpointing Soviet radars so as to determine better ingress/egress routes for our B-52's, in an effort to better avoid their missiles and fighters.  The RC-135's also carried Air Force linguists on-aboard, fluent in Russian, to detect and translate important voice communication, for both real-time information, analysis, and contingency planning use.

While we were conducting continuous surveillance flights very near the Soviet Union, with increased intensity during the Crisis, especially, perhaps not surprisingly, the Soviets were doing some of the same, sending their long-range aircraft up and down our East Coast shoreline, from the legal vantage point of international Atlantic waters.  But, as Shuler remembers, while they were, indeed, doing reconnaissance, in his view, what they were doing was nowhere near as comprehensive as SAC and DoD's plans and missions designed to carry out our retaliatory response, aimed at the "complete destruction of the Soviet Union."

As mentioned earlier, the nuclear weapon load per bomber, during Shuler's time piloting B-52's, was two. Later on, it was possible for each aircraft to deliver four such weapons. "They were loaded onto a clip which fit into the bomb-bay. Those weapons were each in the 1-mega-ton range (vs. the earlier 3-to-9-mega-ton range)," he remembered.

Shuler recalled and mentioned, again, that some of the bombers, in other units, carried the 'Hound Dog' missile. These were nuclear-tipped, two per bomber, and carried externally on the weapon hard-points under the B-52 wings. Many of those aircraft also carried nuclear bombs internally. As mentioned earlier, SAC's operational plan for those dual-capability bombers was to 'shoot-and-penetrate.' "Those guys would pound their way in, using the missiles to take out targets that might have been able to take them, or other of our aircraft out, then continue on to deliver their internally-carried weapons," said Shuler.

These dual-equipped B-52's were based mostly in the northern sector of the U.S., reducing their flight time to the Soviet Union. The plan was for them to go in after our ICBM's, and before our bombs-only aircraft, like Shuler's, which were stationed then mainly in the southern states, like Texas.

"Once we got our ICBM's up, with their fly-out time to target of 30-to-35 minutes (from continental U.S. to Soviet Union/Warsaw Pact nations), with their mission to take-out major enemy targets <u>before</u> launching our follow-on B-52/B-47 strike-force, those ICBM's would become an important, timely, retaliatory, return first-punch, since the flight time to target for our bombers was roughly 12-to-14-hours before we actually delivered our weapons," said Shuler.

By the late 1960's, as a result of the comprehensive Soviet anti-aircraft defensive build up, the B-52 was "no longer a viable penetrator, so we starting producing stand-off weapons, like the air-launched cruise missiles, as well as bringing the B-1 and B-2 bombers on-board, both of which could far more effectively defeat Soviet radars and surface-to-air missile systems," said Shuler.

All told, there were a total of 744 B-52's built, and, today, just 85 remain in active service (back in 1991, the START Treaty with the Soviets dictated the destruction of the others remaining). Our production of B-52's ended in 1962, making all those produced that year (i.e., the 'newest'!), now 55-years old! During "Operation Desert Storm" (First Gulf War/1991), our B-52's, carrying conventional weapons, dropped 40% of those delivered by air. And finally, believe it or not, our remaining fleet of B-52's is expected to continue serving America's defense needs on into the 2050-decade, which at this point, would seem to be on track to set a single military airframe longevity record!

And for the B-52's human-impact honor roll, Shuler indicated that, from March 1946 (SAC's activation), until SAC's de-activation, on June 1, 1991, over 2,500 combat crew members (B-52, B-47, EC-135, KC-97, KC-135, RC-135, and U-2) lost their lives in either training accidents or in combat. Far beyond the cost of the aircraft destroyed over that 45-year SAC operational period, those combat crew members who perished, in service to our nation, became the real toll, the real price paid, for the defense of America throughout that time of unremitting challenge and tension.

A time and tension sadly and, frankly, now either forgotten, or never even known, by far too many of our citizens, who were never factually instructed. Those (especially in positions of national authority) who remain ignorant of our past, through negligence or by design, are destined to repeat it, as wise men have sagely told us. And in all likelihood, America (and Americans) could well be seriously harmed, if not destroyed, by this now-rampant lack of awareness of our past world of nuclear threat and tension. There was simply far too much at stake back then, just as there is today, for the sake and survival of both our nation and precious freedoms, to remain blissfully complacent or ignorant.

Returning again to that Post-Cuban Crisis/Cold War era, along with all of that full-time ground-alert and air operational status during his two-year+ assignment at Dyess AFB, Captain Shuler had somehow managed to wedge-in graduate engineering coursework, under the auspices of the Air Force Institute of Technology (he began his Air

Force career with a civil engineering undergraduate degree).  Then, in
1966, he was selected to enter the Master of Science degree program
in Engineering Management at Rensselaer Polytechnic Institute in
Troy, New York.  There, despite the demands of full-time graduate
studies at a prestigious engineering school, he was still required to
maintain his flight proficiency, so he managed to complete those
needed air-hours, flying the U-3A twin Cessna out of Stewart AFB,
near Newburgh, New York.

After earning his master's degree from RPI in 1967, Shuler received a
directed-duty assignment to the civil engineering career field, which
was intended to be a three-year tour (selection for one-year of
graduate school obligated the officer-graduate to three-years of duty
in that field).

That said, why then, after being selected for advanced professional
study, was his civil engineering assignment abruptly halted and put on
hold?  And, why in its place would this B-52 aircraft commander,
with 2,500-hours of multi-engine military flying time behind him,
receive orders to transition instead, and unexpectedly, to F-4 fighter-
attack pilot training?  The answer, as it is with all unexpected
transfers, proved to be quite simple. As those who have served know
well, the <u>needs</u> of the service branch, in his case, the Air Force,
prevailed.

And here was the need:  At that time, the Summer of 1967, the war in
Southeast Asia was heating up.  By then, all of the Air Force's
existing tactical fighter pilots had already served a combat tour in that
region.  "So, they made a rule that the rest of us multi-engine pilots,
from the Military Airlift Command, from SAC, and guys from Air
Training Command, who hadn't been to Southeast Asia, had to go
there, before they would send our veteran tactical fighter pilots back
for a second tour," recalled Shuler. "I had not been over there yet, so
they said I had to do a combat tour before entering the civil
engineering career field."

Also at that time, the Air Force had a 'Choose Your Weapons"
program, allowing pilots to volunteer for a specific type of
fighter.  Shuler elected to go with the F-4.  A world of difference in

all ways from the B-52, but he had, of course, actually flown a single-engine military aircraft (T-33) as a part of his initial pilot training, so the transition wasn't as stark for him as might be expected. Especially since he would start out his re-familiarization by flying an AT-33, a trainer-modification now equipped with 50-cal. guns.

To do that, he was assigned to the 68th Tactical Fighter Squadron at George AFB in California, as a replacement training-unit fighter pilot in the F-4D Phantom II. There he underwent extensive training in fighter aircraft requirements and tactics, with both flying time (15-hours in the AT-33, plus 6-months in the F-4), along with classroom instruction and evaluations. Of significant training value for Shuler, his instructor pilot was on an exchange program from the Navy, and one who had already flown combat in Vietnam in the F-8! Training modules and practice in California included: air refueling, lots of air-to-ground work (dive-bombing, skip-bombing, rocketry, and strafing), and low-level navigation. "We got to fire the Sparrow missile (radar-guided anti-aircraft). The F-4 carried four 'Sparrows,' and depending on the configuration and mission requirement, we'd also carry four 'Sidewinder' missiles (heat-seeking air-to-air), as well. The Sparrows always remained on the aircraft, regardless of the mission we were flying," recalled Shuler.

Following fighter pilot graduation, he went to Tactical Sea Survival School in Homestead, Florida ("where you learned how to parachute into a water environment and survive!"). Then it was on to Jungle Survival Training ('Snake School') at Clark AFB in the Philippines. "Along with collateral classroom work, we went out on escape and evasion training (day/night), up in the hills beyond the base. There, native youth were used to track down the hiding U.S. aviators. For every one found, the young boys were given a bag of rice," remembered Shuler. As it turned out, finding the 'hidden' Americans in the jungle proved to be a lot easier than the pilots had assumed. The boys had quickly learned to simply follow the tracks left by the larger American boots! When asked if the youngsters had been able to find him and his fellow students, Shuler replied: "Yep, they caught every damn one of us!"

Among other events in that escape training sequence was a very important one, but one these Vietnam-bound pilots hoped they'd never need to experience: extraction following a bail-out. "We practiced being hoisted out of the jungle on a 'tree penetrator.' As the helicopter hovered overhead, they'd drop the 'penetrator' down and we'd strap onto it, as they pulled us out of there. We practiced all manner of rescue operations," Shuler remembered.

In that particular fighter training program with Shuler was fellow Citadel graduate (Class of 1951), Rudy Nunn, who went on to fly Skyraiders in Vietnam, based in Laos. Within just 30-days after his graduation with Shuler from "Snake School," while flying a combat mission in an A-1E aircraft, Nunn was shot down and killed. A stark memory of a friend, colleague, and fellow pilot who lost his life in defense of freedom, not unlike so many similar military member remembrances of comrades lost in combat, from all past wars right through to present day Iraq and Afghanistan, and perhaps other locations yet to come.

Upon final completion of his transitional flight, sea survival, and evasion/rescue training, in March of 1968, Shuler transferred to the 558th Tactical Fighter Squadron, Cam Ranh Bay Air Base, South Vietnam, to begin his year-long tour as an F-4 aircraft command pilot. He arrived first at Tan Son Nhut Air Base (7th Air Force HQ), at the same time his brother Jake was stationed there. "My parents were not too happy about having both of us over there at the same time. We had about a six-month overlap. Our Dad had spent two-years in the Pacific during World War II, so Mother felt like she had already contributed enough!" he recalled.

Once at Cam Ranh Bay, the "check-out" required of newly arriving pilots was flying five missions. "The first mission you flew in the backseat with an instructor pilot, kind of an orientation," remembered Shuler. "From then on, you flew in the front seat, and were married-up with your 'GIB' (back seater, known among Phantom pilots simply as, the 'Guy in Back'!). In this case, it was Lieutenant Bill Reed. I flew with a lot of different GIB's, but Bill and I flew the majority of our combat missions together (Sadly, Reed was killed on his second combat tour, while flying over Laos in an F-4D).

The fifth and final check-out mission was a status evaluation flight, with a supervising officer in the backseat, just to see how you were doing," said Shuler.

The primary mission of the 12th Tactical Fighter Wing at Cam Ranh Bay was close-air support for the Army and Marines on the ground. A secondary mission was enemy supply interdiction on the Ho Chi Minh Trail and elsewhere. Shuler's F-4C carried a variety of ordinance: Napalm; or bombs of varying application and destructive power (500-lb, 750-lb, and/or 2,000-lb). Some of the bombs were 'slicks,' while others were 'high-drag,' released at low altitude, around 450-to-500-feet, at an aircraft speed of about 450-to-500-knots. With the 'high drag,' fins would pop out, retarding the fall of the bomb, allowing time for the aircraft to exit the target area before ground impact and detonation.

The Phantom F-4C also had an exterior 20-mm cannon attached to the centerline of the aircraft (later on, the F-4E was equipped, instead, with an internal cannon), capable of a firing speed of 6,000-rounds-per-minute. The typical mission load carried was around 1,200-rounds, allowing for sufficient short, pulverizing bursts, making the Phantom's presence abundantly clear!

Back in January of 1968, following a massive enemy artillery bombardment poured down upon the 5,000 U.S. Marines garrisoned at the Khe Sanh base in northwest South Vietnam (near the border with Laos), a huge wave of North Vietnamese soldiers then pushed forward to surround our troops. The lengthy engagement (January-July, 1968) that followed is considered to have been one of the bloodiest battles of the war.

"The first operation I became involved with in-country was the siege of Khe Sanh," related Shuler. "I flew six missions in support of our Marines there. Called 'Operation Niagara,' which was appropriate campaign name, because of our heavy, continually flowing tactical air strikes to neutralize the targets that had Khe Sanh surrounded, with air support from B-52's, as well." Of necessity, many of those air-strikes came very close to our Marines and their base's fortified perimeter. In the midst of the surrounding chaos, continual re-supply of the compound was required. That was achieved, primarily, and continually, by Air Force C-130's delivering ammunition, food, water, etc. Given the heavy ground fire from the North Vietnamese surrounding the base, they weren't able to land as normal. Instead, the C-130's came in high, then dropped down over the compound's runway, pushing out drogue-chuted supply pallets in-flight, before quickly regaining altitude to safely depart the area. "We finally broke the siege with 'Operation Pegasus.' Lord only knows how many of the enemy we killed, but they finally quit. Then the Marines could get out of the confines of the base and mop up the rest of it," Shuler recalled.

"I also flew several missions interdicting roads and ferry crossings in Laos. Then I flew another six missions interdicting and hitting bridges in the panhandle of North Vietnam. Otherwise, we were primarily close air support. And the most rewarding missions were when we had troops in contact (with enemy forces) and were able to go in and help relieve those conditions. A lot of missions were termed 'jungle busting,'" said Shuler. Meaning putting concentrated airpower on valued targets around the clock (sarcastically referred to by pilots as "knocking down trees and killing monkeys").

"We also flew missions we called 'sky spot,' which is where you bomb from altitude in level flight, weather permitting. In cases where it didn't, rather than scoring our own bomb hits, as had been done previously, several of our ground radars sites were moved into South Vietnam. That way, putting them closer to the fighting, our radar guys were actually locating the intended targets, with us on their screens as well, and then directing us in, at the right altitude, and airspeed, with the right weaponry. We simply followed their instructions in-flight, made any corrections, left or right, then flew down to the mission point. It became so precise; they would even tell us when to release our bombs! We used this radar-directed method when weather conditions wouldn't permit us to strike visually," said Shuler.

Shuler's initial Vietnam deployment assignment had been with the 558th Fighter Squadron of the 12th Tactical Fighter Wing. But when he arrived at Cam Ranh Bay, of all things, the 558th was in

Korea! This was the time of the unprovoked capture-at-sea of our intelligence-gathering vessel, the "USS Pueblo," by the North Koreans, sparking a major crisis. Thus, more tactical air assets were required to bolster our response capability in Korea. Although he would've preferred otherwise, the Air Force then 'suggested' that he needed to join them. Before temporarily departing Vietnam (April, 1968), just a month after his arrival, as indicated, he had already flown his first in-country combat missions (a total of seventeen) with the 391$^{st}$ Squadron, which quite effectively "got my feet wet," he recalled.

"Our primary mission in Korea was flying 'MIG Cap' for the RC-130 reconnaissance flights, which meant flying up both coasts of North Korea and along the DMZ (Demilitarized Zone, separating the two Koreas). We'd fly in a two-ship compliment (F-4's), weaving over the C-130's, with our 200-mile-range radars, looking into North Korea to detect any MIG's that might be heading to try to knock down our '130' platforms," said Shuler.

While he was there, but not in the air, the North Koreans did shoot down one of our helicopters along the DMZ. Then, after he left, they shot down a Navy RC-121 reconnaissance aircraft (with the loss of all personnel onboard), and another helicopter. So, needless to say, those protective RC-130 'MIG Cap' flights were both important and dangerous. Shuler ended up flying 15 of those combat support missions in Korea, along with 57 regular training sorties, to include qualification for nuclear weapons delivery with the F-4.

After five months in Korea, a U.S. fighter squadron stationed in Japan came in to relieve the 558$^{th}$ so they could return to the fight in Vietnam (September 1968). "During that time, I had become 'lead-qualified,' both for two and four-ship flights. Those first 17 missions in Vietnam, I flew as a wingman. So that when I got back to Cam Ranh Bay, I could then lead missions. Out of the total missions I flew, I led about a third of them. And not everyone became lead-qualified. I had lieutenant-colonels flying my wing who never became flight leaders," recalled Shuler.

Shuler flew missions within all four 'corps' of South Vietnam, as well as up North, some with our troops in enemy contact, but most of the time, not. "If we were going on an interdiction mission to North Vietnam or Laos," he remembered, "we'd generally take a 'four-ship' (aircraft) flight, but due to the distance, and carrying a full ordnance load, that always required us to refuel between Pleiku and Da Nang, with a KC-135 tanker aircraft. Most of the 'in-country' missions were 'two-ship'. If we were going against fortified emplacements, you'd have one aircraft with 500 or 750-lb. 'high-drags,' and the other guy would have wall-to-wall napalm. So we'd go in, the lead with the 'high-drags' to open up the bunkers. Then the second aircraft would come in behind us and pour the napalm," said Shuler.

"We also flew 'top cover,' if you will, for the C-123 'ranch-hand' mission. That's where they (the 123's) would fly low to put out the jungle defoliant, via spray-bars on the wings (killing the trees and leaves). We would fly over them, and then down in front of them, to try to draw ground-fire, keeping one F-4 up on the 'perch.' If we did get ground-fire, then the guy on the perch was ready to go in and strike the area."

Shuler then recounted a couple of his other missions, specific, memorable ones that he referred to as 'interesting'! "And let me say, first, I was fortunate. I got shot at a few times, but I was never hit. I was damn lucky!" By contrast, his friend, Major Don Lyon, was flying in an F-4 over Laos, on his very first combat mission in the back seat. His aircraft was shot down and Lyon was killed. First mission!

Thinking back to the high rate of American bomber crew losses, most pronounced in the early months and years of World War II, Shuler's "damn lucky" rings fatefully true, even to present day combat, and, regrettably, sometimes with training flights, as well. Those 'interesting' flights he spoke of, occurred while serving in Vietnam. During one such, he was leading a two-ship mission over to the tri-border area (Laos/Cambodia/South Vietnam). That area was considered to be a very fluid border confluence, with North Vietnamese troops frequently coming through it.

On this particular interdiction mission, said Shuler, "I made contact with my Forward Air Controller (FAC). Anytime we struck in South Vietnam, we always had a Forward Air Controller. There were no free-fire zones. The Air Force FAC, flying an L-19 "Bird Dog" or an O-2-B Cessna, would mark the target with a smoke rocket, then give us directions where best to drop the ordnance." The FAC then told him to stand-by, while coordination was clarified with a couple of Army FAC's also flying over the designated target area, since there were U.S. troops in the vicinity, as well as enemy. "Then, the next think I heard over the radio, recalled Shuler, was, "Oh, my God!" Those two Army FAC's had collided in the air and gone down. A helicopter was quickly summoned, and a rescue team inserted. Regrettably, one pilot was killed; the other, severely injured (he was immediately extracted and flown back for treatment).

With the attack delayed due to the accident, Shuler and his wingman remained overhead in the area until their fuel level approached "bingo" state (i.e., needing to depart and refuel). The Air Force FAC then radioed Shuler: "OK, if you've got to leave. I'm going to show you the target area, and I just want you to dump your whole load! So, he fired a smoke rocket in there. Then I rolled in, released everything, 'pickled' all the bombs off, and pulled out. Then, I heard the FAC on the radio yelling: 'They're shooting at you. They're shooting at you!!' My wingman got his bombs off OK, and then we beat feet back to Cam Ranh Bay." Shuler recalled that, at times, he could actually see the ground fire, especially in Laos and North Vietnam (heavier guns).

On another 'interesting' mission, this time carrying, and then dropping, napalm over the target area, his wingman, as was customary, had done a visual check of Shuler's jet, and immediately radioed him that a 'can' had gotten hung-up under his aircraft. That apparently had resulted when the forward 'lug' had released, but the rear one had not, causing the napalm 'can' to twist in the slipstream, and in short order was punctured by the 'sway brace," creating a very serious situation. "All that jelly gasoline was siphoned out and, as his wingman had indicated, was sticking to the bottom of the airplane, practically covering it," remembered Shuler.

With his wingman alongside, but now at a safe distance, Shuler returned at once back to Cam Ranh, declared an emergency, flew a straight-in approach, and landed, fire trucks on either side, as he brought the aircraft to a halt and quickly shut the engines down! He and his rear-seater exited the aircraft and then got their first real look at their plane, coated in napalm. "I don't know that anything would've set it off," said Shuler. "But had it torched, we'd probably have been crispy critters!"

When asked what could have set it off, he shared a story from that time and locale about his brother, Jake Shuler, who was an F-105 pilot. That particular day, Jake was in a 16-aircraft strike-force flying up to the Hanoi area. As they approached, the enemy began shooting SAM missiles at them. So, the flight leader called for a 'SAM break,' meaning the pilots were to hit their after-burners, immediately push over and head down to the ground as close as they could safely get. The F-105's had an external fuel drain-line (an overflow spout) on their tanks, sending any excess out the back end of the aircraft. When his brother hit his 'burners' to rapidly accelerate, and swung hard toward the ground, the draining fuel, at the rear of his aircraft, burst into flames. His wingman looked over and saw a huge ball of fire and called for Jake to bail-out. Jake immediately checked his gauges and all readings appeared normal. He then checked his mirror and could see the fire behind him, but decided to keep going, telling his brother later that "he wasn't about to bail-out over downtown Hanoi!" Fortunately, as he pulled out of his dive, no longer needed for extra thrust, the afterburners shut off, and the fire went out on its own! A safe ending to a tense event for brother Jake. And an example of what could have happened to set off that napalm which had coated the entire underside of Shuler's aircraft.

Continuing to reflect back, he remembered that, at that time, we were actually bombing in Laos, a fact closely held, although air strikes there were authorized. "The rules of engagement were rather stringent," remembered Shuler. We had written tests, monthly, on rules of engagement, what you could and couldn't do. It was pretty serious business. Also, we were not allowed to fly into Cambodia, even thought that border was very porous," he said. "That's where the Ho-Chi-Minh Trail came into South Vietnam."

"Then, in October of 1968, we mounted what was probably the most massive tactical air campaign up to that time," said Shuler. "Seventh Air Force headquarters decided that we were going to put a stop to that Ho Chi Minh Trail traffic. So, the Air Force picked out the critical junctions, and maintained tactical air over those points 24-hours a day. We bombed that place, those nodes, to the point where, as we found out later, it became so tough for the North Vietnam Army troops (NVA), that they had to withdraw three divisions, including some Viet Cong, because they couldn't re-supply them, they couldn't feed them, they couldn't arm them. We definitely put a stop to it all. We really had them pinned down," recalled Shuler. The principal objective had been to render the "Trail" impassable, effectively collapsing it. Non-stop Air Force bombing had done so. Said Shuler: "We had used some unique ordinance to do that. We carried the CBU-49's, a bomb-type that opens up after release and spreads little bomblets all over the place. They're time delayed. You don't know when they're going off. So, we'd put that on the roadways, nodes, and junctions. Beyond road damage, they were very effective against personnel and trucks, as well."

"Then, in October of 1968, we mounted what was probably the most massive tactical air campaign up to that time," said Shuler. "Seventh Air Force headquarters decided that we were going to put a stop to that Ho Chi Minh Trail traffic. So, the Air Force picked out the critical junctions, and maintained tactical air over those points 24-hours a day. We bombed that place, those nodes, to the point where, as we found out later, it became so tough for the North Vietnam Army troops (NVA), that they had to withdraw three divisions, including some Viet Cong, because they couldn't re-supply them, they couldn't feed them, they couldn't arm them. We definitely put a stop to it all. We really had them pinned down," recalled Shuler. The principal objective had been to render the "Trail" impassable, effectively collapsing it. Non-stop Air Force bombing had done so. Said Shuler: "We had used some unique ordinance to do that. We carried the CBU-49's, a bomb-type that opens up after release and spreads little bomblets all over the place. They're time delayed. You don't know when they're going off. So, we'd put that on the roadways, nodes, and junctions. Beyond road damage, they were very effective against personnel and trucks, as well."

"The other thing we carried was something called BLU-52's. That was a finned, napalm canister, that had a 'CS-agent' in it. Again, on the roads, and all, that thing would open up and spread this dust all over the place, causing enemy troops to vomit, impair them mentally, and basically lose all bodily control. One of my colleagues reported to me that he watched an enemy bulldozer driver drive off a damn cliff. Obviously, he'd gotten into some of that stuff!"

*Capt. Buck Shuler with his F-4C, named after his wife, "Annette", January 1969 at Cam Ranh Bay Air Base, South Vietnam*

Overall, by employing the special ordinance elements he outlined, the results proved to be very effective, the mission deemed a success. "At the very least, it made their (enemy) lives miserable," recalled Shuler.

Given the impressive results of superior American air power, as in the Ho-Chi-Minh Trail example, our bombing campaigns were having a definite negative impact on enemy intentions, and especially on their ground forces. No question, then, that the one thing the enemy most needed, at that critical point in time, was a bombing-halt. And, regrettably for our war fighters, and especially for all those who'd sacrificed their lives, American politics, starting with the President, over-rode mission-sense, and promptly answered their wish.

"The last day of October, I flew a mission," said Shuler. "That night, at mid-night or thereabouts, LBJ (President Lyndon B. Johnson) called a halt to all bombing in North Vietnam! Later, I saw reconnaissance photos of the Ho Chi Minh Trail, and I'll tell you, it looked like a busy interstate. They were able to come in there, with heavy equipment, and repair all of the bomb damage, bring the roads back up to speed, and the trucks rolled, and rolled and rolled" said Shuler, with no little disgust in his voice, the memory still vivid, even to this day.

The U.S. had experimented at least once earlier with bombing halts in the north. And even prior to that, Air Force aircraft had actually been forbidden to strike any airfields in the Hanoi-Haiphong area, or any other targets within that entire zone. Even though the North Vietnamese Air Force had Mig-21's lined up in plain sight on base ramps, almost as if to taunt. Regardless, American aircraft were not permitted to strike them, a source of major frustration, to put it mildly, since parked down there before them, were the very enemy aircraft being used to shoot down American pilots.

"Later, "recalled Shuler, "after I came home, they (U.S. Air Force) initiated operations named 'Linebacker One' (May-October, 1972), and 'Linebacker Two' (December, 1972), with 'Two' being the so-called '11-Day War,' where Nixon (President Richard Nixon) unleashed the B-52's against both Hanoi and Haiphong. That's when we had them hanging on the ropes for sure. We literally ran them out of surface-to-air missiles. And our planes shot everything they had. So that, by about the tenth day or so, we could fly up there with impunity. We did, however, end up losing some B-52's (about 15 total during the entire war)" he said.

Going back to the tail end of his Vietnam tour, at Cam Ranh Bay, by January of 1969, even then, Shuler could tell the war was beginning to wind down. One telling sign: The Wing began a base beautification project! Among other things, they started planting palm trees along the base avenue. And the Wing Commander made the decision that flight suits could only be worn by personnel scheduled to fly; otherwise, khakis. That "didn't sit well with a bunch of damn fighter pilots," recalled Shuler.

Aircraft attrition there also became a sign that our participation was becoming less intense. "By the time I left the 558th, they had only 11 airplanes when they normally would've had between 18 and 24," he remembered. And that reduced number of aircraft made it even more difficult to get on the flying schedule. "My philosophy was, if you're going to be there, I'd just as soon fly every day. I'd flown as many as three combat sorties in a single day. Now, it became tougher and tougher to even get on the schedule."

"In the year I was there (1968-69), we lost ten pilots and sixteen airplanes, two of which crash-landed at Cam Ranh Bay with battle damage. But our loss rate was far less than those guys flying out of Thailand, way up north into the Hanoi-Haiphong area. The first 51-days my brother Jake was based in Thailand (Khorat Air Base), that Wing, the 388th TAC Fighter Wing and the 355th TAC Fighter Wing (Takli Air Base) combined, lost 48 F-105's, including among those pilots, two full-colonels," said Shuler. All of that in just 51 days!

*Captain Shuler returning from his
combat tour
in Vietnam,
greeted by sons,
young Buck and Frank, Mobile,
Alabama on March 2, 1969.*

All told, then, during the 12-months he was deployed, first to Vietnam, then Korea, and then back to Vietnam, Shuler flew a total of 180 missions (122 combat + 58 training), over North and South Vietnam, and Laos.

"I think most of us (air crews) came back from Southeast Asia with a real sour taste," said Shuler. "I think the consensus was that we had not been permitted to apply airpower as it should've been. For instance, the first build-up began in about 1965. If 'Linebacker Two' had occurred in December of '65, rather than December of '72, no question, the war's outcome may well have been dramatically different." And thousands of American military lives could have been saved.

Shuler was asked what one word might best sum-up his feelings about the Vietnam War. That word: "Disappointed, to put it mildly. If I had to use a stronger word directed at senior leadership, I would say <u>failure</u>, referring mostly to the civilian administration at the time, and to some degree, military leadership, as well, but not nearly as much." Years later, then-Vietnam War-era U.S. Secretary of Defense Robert McNamara freely admitted in his book (<u>The Fog of War</u> / 2003), said Shuler, that we (civilian leadership) made "serious mistakes."

Earlier in the war, "The Joint Chiefs of Staff gave him (President Johnson) a game plan that included a blockade of Haiphong Harbor," said Shuler. "All the things we ended up doing in December of '72, that should've been done much earlier. LBJ was apparently worried about getting into a shooting war with the Chinese, so that plan was never approved when presented by our military leadership."

"And let me tell you the other thing", continued Shuler, "and I only learned about this about a year ago, through documentation. Dean Rusk, the Secretary of State at the time (1961-1969), admitted that, through back-channels, he had actually revealed to the North Vietnamese the targets we were going to hit! Meaning, we now know, that anyone of our guys who flew those missions, 'Rolling Thunder' (March 1965-November 1968) and the like, were then even more at risk." This secret administrative policy, uncovered years later, was apparently aimed at limiting 'collateral damage' (killing Vietnamese civilians). While perhaps laudable from a purely self-serving political posture, it proved to be treacherous (if not traitorous) from the standpoint of our military aviators who then faced even greater danger, and heightened potential for shoot-down, with the enemy knowing our targeting objectives in advance. Hard to believe, and harder to accept, that politics could so callously override the well-being of our nation's war effort, and in particular, of course, the very lives of our warfighters.

But perhaps the Rusk travesty, as ordered by the President he served, shouldn't surprise us in hindsight. History has now shown that, while the American-led forces finally had the North Vietnamese, by their own subsequent admission, 'on the ropes,' with costly, hard-fought victory at hand, it proved to be America's over-ruling political leadership who finally chose to simply conclude that war, rather than win it.

In March 1969, then-Captain Shuler returned to the States, with an assignment to Second Air Force at Barksdale AFB (Louisiana), finally in the well-prepared, long-awaited role of industrial engineer on the Headquarters engineering staff, soon advancing to Deputy Chief of the engineering division. In January 1970, he was promoted to Major.

As he continued to excel in his duties and tours, additional well-earned opportunities and promotions would continually come, culminating in his 1988 assignment as Commander of the legendary 8th Air Force at the rank of Lieutenant-General.

*Lt. General Shuler's mother, Mrs. Berta W. Shuler, and his wife, Annette, help to unfurl his three-star flag at Barksdale AFB, Louisiana, on March 26, 1988, prior to his assumption of command of the 8th Air Force.*

General Shuler retired from the United States Air Force, effective June 1, 1991, completing 32-years of military service, with more than 7,600 flying hours, including 209 of those in combat during the Vietnam War. He spent a total of 23-years serving with the Strategic Air Command, ten of which as a General Officer, flying 358 missions as the Action Officer with the emergency airborne command post (EC-135 "Looking Glass") during the four-decade long Cold War. General Shuler's military awards include the Distinguished Service Medal (with oak leaf cluster), Legion of Merit (with oak leaf cluster),

Distinguished Flying Cross (for actions described previously), the Air Medal (w/five oak leaf clusters), and the Air Force Commendation medal (with oak leaf cluster), along with many other decorations and ribbons.

**UPDATE:** E. G. "Buck" Shuler, Jr. served in several command positions around the nation, following his B-52 and F-4 assignments overseas in both "cold" and hot war scenarios. In the early 1980's, he served as commander of the 3$^{rd}$ Air Division stationed in Guam.

After an assignment at Strategic Air Command Headquarters (1986), LTG Shuler received his third star and, with it, command of the 8$^{th}$ Air Force at Barksdale AFB in 1988. He retired from the U.S. Air Force in 1991. He and his wife, Annette, now reside in Sumter, S.C.

# CHAPTER 14

## "We Took A Pretty Good Bloody Nose When We Got to Ramadi! Colonel Mike Squires, West Point/Ranger United States Army (Ret.)

One memory held by then-Captain, now United States Army Lieutenant-Colonel Michael T. Squires, early in his first Middle East combat tour (2003). "Sadly, those stateside training deficiencies quickly came back to haunt us, once on the ground and in combat in Iraq. Now, I don't remember exactly when it happened, but somewhere between two-to-six weeks into our time of showing up into theater in Ramadi, the NCO's kind of collectively woke up and realized, hey, I am an NCO, I am being empowered, I am needed, I have a role here, and with that realization, they picked up their game," recalled Squires. "And when that happened, the whole unit turned around. From that point, we were running on all cylinders. Once we hit that mark, and, again, it was probably more like six-weeks, then I finally felt comfortable concluding that, hey, I'm on a good team now!" he said.

Looking back to a much younger time in his life, perhaps eighth or ninth grade, Mike Squires had an inkling that military service might be in his future. While several members of his extended family had served (Marines, Navy, Air Force), that really wasn't what initiated his military thoughts. As he recalled, "I don't know if it was the 'Be All You Can Be' TV ads, or maybe the Gulf War, but certainly the connection with being part of a team that was bigger than yourself, and doing something for your country, so it was kind of a two-fold thought."

A start of an idea, perhaps, but nothing that really took hold. That would come later. As a High School student, while attending a Boys

State Conference, Squires met, and was impressed with, a Cadet from the United States Military Academy, who was there as part of a summer assignment. From that conversation, the idea of combining service to the nation with a quality college education came into much sharper focus. Following that chance meeting, knowing very little about West Point, he began researching it, along with the offerings of the other branch academies, given the varied service of family members. And after a lot of digging, and careful consideration, he reached the conclusion that "since this comes with a five-year commitment post-graduation, and I'd way rather do that commitment in the Army than in any of the other services."

Decision made, now all he had to do was get in! Squires sought out the Congressman from his residential district in Cleveland. And after a series of applications and interviews, he fulfilled his first career-path goal, earning an appointment, and admission, to West Point (1994). And thankful he was for both, since, with his heart now set on attending the Academy, he freely admits he had no 'Plan B.'

The transition from high school to West Point, while no doubt a challenge for most first-year students, was effectively seamless for Squires. He went from an early breakfast-school-athletics-dinner-studies-late to bed schedule at home, to basically the same routine, though doubtless more rigorous, at the Academy. "I was much more prepared for West Point, than I originally thought I would be, because of my very structured time through high school," he recalled. Regardless, as would be expected of all dedicated students at this demanding level, the West Point experience did provide a lasting impact. "It took time, a good four years, for it all to sink in. The enormity of the task, the responsibility, and profession of arms, all of that took a while to really take hold. It led me to realize that, hey, this isn't just about college. This is about a much larger, more important task. Even if you were the lowest officer or private in the Army, this experience proves just how key your individual role is to the big team. And that probably sunk-in, fully, by the time I was about to graduate," remembered Squires.

Along with the continuing attention to those bigger-picture Army requirements and expectations, on the academic side, his diligent efforts to achieve there culminated with a very impressive, make that a truly incredible, finish. Out of nearly 1,000 seniors in his West Point graduating class, Cadet Squires graduated 5 in his class!

The first assignment for newly commissioned (May 1998) Infantry Second Lieutenant Mike Squires was the Infantry Officer Basic Course (IOBC) conducted at the "Benning School for Boys" (graduating officer term of endearment, in hindsight, for the then-Infantry Center at Fort Benning, Columbus, Georgia!). Upon course completion, he remained at Benning for both Airborne (Jump) School (3-weeks), and then the start of the demanding two-month/three-distinctive element training sequence required to earn the coveted Ranger Tab.

Why seek Ranger Qualification? "I knew as soon as I branched Infantry, that it was something that was pretty much a requirement. If you want to do well in the Infantry, you'd better be Airborne and Ranger-Qualified, otherwise you're gonna have a hard time making it," said Squires. "That's something you learn over four-years at West Point and something you quickly learn at IOBC, when you look around and see who's Ranger-Qualified and who's not. Even among the cadre of officers and enlisted teaching us, there was a noticeable difference between those who had it, and those who did not."

Lessons learned from the demands of Ranger training? "No doubt what it taught me is that you can keep going much longer and further than you probably would let yourself," recalled Squires. "And you know a big part of that is that there's a team with you, and you're not gonna let that team down. Whereas you might think you're exhausted, you might think you can't take another step, but if the guy in front of me is, and the guy behind me is, then I can keep going too. That has served me well going forward. The fact that, if you made it through Ranger school, you can handle whatever future tasks you'll face."

Following preliminary Ranger orientation at Fort Benning, students then face the challenges of conducting operations in the mountains of North Georgia, followed by the swamplands of Central Florida. When asked which of the three sequences presented the most vivid memories, there was no hesitation in his response. "The mountains! Now, I won't say the Fort Benning phase (the first of the three) was easy," he said. "There were a lot of 'go-no-go' tests. Miss this even by one push-up, see you later; miss another event by one-second, see you later. But once we got into the actual patrolling, it became easier. It was squad-level, small. I was familiar with the terrain, the weather was still decent, it was Fort Benning. But, then, we moved up into the mountains, at the end of February, that was a tough transition. The terrain was much more challenging, with a lot of elevation. The moves felt much longer, the weather was rough. One of my 'favorite' memories was when we started up Mount Yona, a famous mountain all Ranger candidates must climb there. It was a cold rain at the bottom, and half-way up, it was a cold snow, to give you some idea of the elevation we gained" said Squires, feeling that cold all over again!

He did recall, and credit, the genuine concern shown by the instructors if any of the students got wet. "We had to cross a creek, and everything was going fine, until my Ranger-Buddy slipped and fell in, got completely soaking wet, and it's about 2 AM," he remembered. The training cadre reacted immediately, realizing the dangerous combination of cold and wet. They literally stopped the whole patrol and formed a perimeter around that soldier. They got his wet clothes off, wrapped him in towels, while others pulled a fresh uniform and boots out of his backpack, and got him dried off and relatively warm, as quickly as possible. "Those sequences in the mountains were tough," said Squires, clearly thankful that phase was now but a memory. Reliving those winter-time North Georgia experiences helped him recall how comparatively easier the following movements were in Florida (training phase three). While they were longer, they were flat! "But I remember my body having so little left (coming off the mountain phase), that I knew I should be warm, but I was cold" remembered Squires. "So even when the sun would come up, and by then it's a beautiful April day, the kind of day I really appreciate now, but back then, my body just couldn't warm up, until we were a few miles into our march."

Lieutenant Squires successfully completed Ranger school, graduating and with his Tab now proudly in place. He, then, headed for Vicenza, Italy to join what would soon become the 173$^{rd}$ Airborne Brigade, his first operational unit assignment. His very high West

Point class rank had enabled him to pick both his branch and his first duty-station. "That was one of the huge benefits of doing well academically," he said. "And, again, I approached West Point like I did high school. At both, I had a series of tasks, so I put my head down, and didn't stop until they were completed. Going back years later to teach at West Point, he realized how much harder studying had become, now, with social media and other distractions that simply did not exist, or were in their infancy, when he was an undergrad.

So why choose Italy for a first post-schooling assignment when, due to his high West Point senior class ranking (5th), he had his choice of any location, across the U.S. or overseas? "There were all kinds of nice things about the unit that made me choose it," said Squires. "It was an airborne assignment. And every officer there was Ranger-Qualified. And, of course, it had the benefit of being in Italy!" As pleasant it may be for lifestyle, however, there is no real large-scale training capability for our troops in Italy. That necessitates two major, extended, training trips a year to Germany for the 173rd.

What he didn't fully appreciate about that, until moving on to later assignments, was the value of the extensive training exercises and small unit cohesion made possible by these annual German field opportunities. "Everyone who goes up there, from Day 1, starts the progression from the smallest marksmanship training, all the way to the highest collective training you can get. And you're there for every repetition. So that team is doing rep, after rep, after rep. The exact same team, together, throughout that process. In sum, two key advantages as I look back: Everyone was there as a team, together from the start, and the NCO's there were very strong" he recalled.

In a later assignment as a company commander, those cohesion advantages Captain Squires experienced with the German training modules looked all the more valuable to him, since he was unable to replicate them here. "I then had a company of 140-150 soldiers, and at any given time, when I'd go to the field, I'd only have 80-110 of them. And it wasn't always the same 80-110! So, I didn't have that team going through every step together. So, although we might finish the same training density, the same marksmanship from team to squad, on up to platoon maneuvers and live-fires, when you look at the proficiency on the backside, my elements out of Italy were several steps ahead of those I trained at Fort Riley years later," felt Squires.

He was dispatched from Italy to Kosovo three times while serving with the 173rd. "One of those times, I went as the battalion's S-3 air officer. I would plan all of the air operations, so on this one, I went over early to set-up the drop zones with a couple of NCOs. More than anything, this rotation was meant as a show of force, demonstrating that the U.S. Army could push a battalion worth of folks to Kosovo relatively rapidly. The second time I went was for a real short planning conference. And the third time I went as the battalion support platoon leader, in charge of getting all of the air and ground vehicles sync'd up," recalled Squires. All his trips were operational assignments, but none of them involved combat. The fighting was pretty much over by that time. "We were there basically serving as a roadblock, keeping people that hated each other apart. We had live ammunition, but I don't think anyone fired a single round. If they did, it was an accident, and they got fired for it," he said.

As a footnote, there's no doubt one more reason for his very positive feelings and experiences recalled about Italy. While stationed there, fortuitously, he met a young American woman who was traveling through Europe that summer. The attraction was immediate. They stayed in touch, she back in the states, and later they spent more time together in Italy. Although Lady Luck's dice certainly played a role in their what-were-the-chances first meeting, this proved to be far from a gamble, but a rock-solid relationship meant to be. Mike and Sarah were married, and now have two great kids, rounding out their terrific Army family.

Upon completing his three-year tour with the 173rd, he went back to Fort Benning to take and complete the Infantry Captain's Career Course, and two other training sequences (one, a familiarization with the Bradley Fighting Vehicle, since he'd next be maneuvering with them) before heading to the Army's First Infantry Division at Fort Riley (Kansas). There, he initially spent time as an assistant operations officer (S-3).

Before long, his unit received a 'no-notice' deployment order. It had become apparent to our military leadership that the situation in Iraq

($2^{nd}$ Gulf War, Iraq Invasion, 2003) would take more than the Third Infantry Division and the $101^{st}$ Airborne Division to get it under control, meaning Squires' brigade at Fort Riley would be heading down-range as well. So, in September, 2003, he was off to Iraq, landing first in Kuwait, before the brigade linked-up with their armored vehicles, and then convoyed with them up to Ramadi in Iraq. This deployment would last a full year. "I did a few different jobs during that tour. Predominantly, I was a plans officer in the 3-shop, and then I did a short stint as the battalion's logistics officer (S-4), and then, kinda last minute, replaced a company commander who was relieved, so I took command overseas of a mechanized infantry company," recalled Squires. "Thinking back, it was an interesting dynamic going from Italy, where every officer was Ranger-Qualified, as were the bulk of the NCOs, then to Fort Riley where that was not the case."

Back when he first arrived at Fort Riley, he discovered that there were only five Ranger Tabs in his entire battalion, and Captain Squires was one of the five! And all five were officers serving in positions of battalion staff and leadership. He realized that the Tab was not the be-all end-all, but he harkened back to one of the valuable lessons he'd learned from his Ranger experience, proving to oneself that, when put to the test, you can do more than you thought you could.

"As such, I was very nervous when we got the call to go to war, because I was not impressed with the (stateside) training level of my battalion. I'd seen the complexity of the live fires that we'd done, particularly the dismounted live fires, compared to what we did in Italy, and they were not nearly as well versed (i.e., skilled)," said Squires.

"I've tried to go back in time to figure out how we, as officers, failed them (NCO's) during the garrison training time. Obviously, officers had failed to school them in basic combat behavior, their responsibilities, and the level of oversight expected, that good NCOs always achieve, but qualities they either didn't have, or failed to exhibit, when I got on the ground at Fort Riley," he remembered. And he also remembered quite clearly, by stark comparison, that those NCO leadership qualities had been ingrained and consistently projected when he was involved with field training with the 173rd back in Italy.

Thinking about the obvious, and dangerous, execution dichotomy, Squires had come to the only plausible explanation for the Riley NCO's expectation and performance gap. "It was likely rooted in the fact that a mechanized unit, such as theirs (First Division), is so platform-centric, so focused on being able to keep the vehicles (Bradley Fighting Vehicles) running, and to be able to shoot from and maneuver them, that all the focus was there, and not necessarily on the people. We (officers) had just become too platform focused in our approach to training, so the problem-sets we gave the team (NCOs) failed to properly address the soldier side of the equation," concluded Squires.

Moving back, now, to the Iraq deployment (2003-04), as then-Captain Squires experienced it, and his impressions of that nation, after spending time in Afghanistan, as well. "One of the nice things about Iraq, that I saw, was they had a good 'skeleton' there. Meaning there was capability there, infrastructure-wise, to actually support and function as a country," he recalled. "They weren't where they needed to be, but they had potential. Driving, as I did, through Baghdad, over to Ramadi, to Fallujah, and seeing some of the other cities, it became obvious to me that there was some brainpower, some planning, some capability there. Obviously not fully developed, but I wasn't overly concerned about the long-term health of Iraq, at that time. In my mind, they would figure it all out and eventually get there. In hindsight, I did think that would've happened five-years after I was on the ground in Iraq, not over twenty-years later," he said.

That delay in becoming the secure and maturing nation he envisioned was, of course, caused almost solely by the political decision in Washington to pull all of our troops out of Iraq prior to the 2012 U.S. Presidential election, creating a void quickly filled there, in the worst of ways, by a destructive force known as ISIS.

Back on the combat side in Iraq, Squires' group experienced a rude awakening once there on the ground. "Those first six-weeks were pretty tough. My battalion (First Battalion/16th Infantry Regiment/First Brigade/First Infantry Division) suffered a total of sixteen soldiers killed in action, over the course of that deployment. Several of them early on, and several really strong leaders were injured in early fighting. A really influential platoon sergeant lost his eye and had to go back. One of the platoon leaders in the company I ended up taking over got his leg blown off. Overall, there were just a lot of casualties in those first few weeks," Squires remembered.

Focusing on those 'rough' first few weeks in country, aside from the obvious danger accompanying missions outside the 'wire' in an active combat zone, the Ramadi headquarters compound (an old, large, multi-acre, former Iraqi Army air defense facility) where Captain Squires' company, his entire battalion, and a total U.S. force of 2,500+ soldiers were stationed, ironically proved to be, itself, a targeted source of danger. "We probably got some kind of indirect enemy fire on our camp, at least every other day, over the course of that year. Usually it was just a nuisance, but every once in a while, they'd do some real damage," he remembered.

The unsettling sound of nearby gunfire was a daily occurrence around Ramadi. Not necessarily involving Squires' unit, but sometimes they did involve his battalion. "Some were small, little gun fights, but then some were horrific, longer-term, big-time fights, like the one in which that platoon leader from Charlie Company lost his leg," he said. Worse yet, our soldiers usually couldn't predict when, and from where, these attacks were coming. The enemy had learned the hard way early on that they simply could not take on U.S. troops directly, since, thankfully, our forces had too much firepower and too many ground troops along with it. To help overcome that allied advantage, the enemy resorted to the prevalent use of IED's, the elusive weapon of choice still employed by the opposition on the battlefields of today.

Recalled Squires: "This was a fuzzy time (in Iraq), because you didn't know who in the neighborhood was Al-Qaida, or in this predominantly Sunni area, whether it was just their tribes fighting against each other, or even perhaps Sunnis fighting against Shia forces, intent on taking over territory in the Sunni's Ramadi stronghold. So that kept our intelligence shop busy trying to figure out who was who, and who or where to target and go after. Unlike most previous wars, it wasn't always clear who your enemy was, until they were shooting at you!"

Squires was asked to reflect back on the first skirmish or battle that he, personally, found himself in during his time in Iraq, now some 14-years ago. "The first big firefight that I was in," he recalled, "was when we did a pretty significant operation along with one of America's elite units. They had flown into our headquarters and said, hey, we've got some bad guys here and we need your help with an outer cordon. So, we planned and rehearsed that nighttime mission, and went in. I was stationed on that outer perimeter, commanding and controlling ('C2') the assets we had manning it. The elite unit thought they had the particular target-house pinpointed. Based on that, my men and I were about three blocks away from the fight … or so we thought!!," said Squires.

The elite's suspected target turned out to be the correct one, and a huge firefight ensued. But, as it happened, that target house fight ended up being very close to the protective outer perimeter, a whole lot closer, in fact, than Squires and his men had anticipated. "So, things got pretty interesting really fast," he recalled.

A perhaps unexpected amount of automatic weapons fire came at the Americans from the enemy structure. As you would hope and expect, we had a powerful response. Backing up our dismounted forces, we had several Bradley Fighting Vehicles pouring massive amounts of ordinance into that targeted building. Despite that, a determined enemy fought on. Regrettably, our elite unit did lose one of its team members during the course of that intense fighting.

At the perimeter, which, more than security, quickly became a part of the fight itself, Captain Squires was coordinating periphery assets from, of all things, an old Korean or Vietnam-era armored troop transport (M-113). It had a 50-cal. machine gun trained on the suspect building, continually firing. Because of it, his location attracted responding fire from the enemy, as did the soldiers around him, manning Bradley Fighting Vehicle guns.

"In this particular instance, we got lucky," recalled Squires. "There was my vehicle, and about twenty meters away, was another one, and two RPGs landed literally right between the vehicles. There's no doubt those Bradleys could've sustained an RPG hit with only minor damage. Had that hit mine, however, I don't know how much damage it would've done. Depends on whether it might've been a really 'good' one, with an armor-piercing round, that likely would've severely injured or killed us. Or simply an old one that maybe, at most, (the jolt) or (the ringing in our ears) would've given us a migraine headache the next day!

This nighttime mission to subdue a stubborn enemy target turned into a three-hour-long battle. Having understandably lost all sense of time, Squires remembers that when his unit finally got the OK to pull away, the sun was just coming up. Said he: "I know our side killed a bunch of them. We had between 25–30 KIA on that target that night. No enemy fighter was left alive, when the compound was searched. So, in my mind, there's a bunch of bad guys who aren't around anymore. And then that particular neighborhood was a little quieter for a month or so. It had been fairly 'active' prior to that." The best news of all was that all of Squires' soldiers survived. He felt it was an effective firefight for them, with some valuable lessons learned.

"So that was my first baptism by fire, if you will. I had already been on a bunch of missions, but none that proved to be that hot," recalled Squires. Beyond the unexpected closeness of that firefight to his perimeter position, there was another reason the memory of that particular mission had managed to stick with him. "I distinctly remember that night was Halloween, and what a strange, unforgettable 'Happy' Halloween experience it became!"

Besides that haunting holiday, it had also been for then-Captain Squires, as mentioned, his first close encounter with an actual combat event. Thinking back, what impressions were made, what feelings does he recall, were going through his mind during that fight? "I've read many books about soldiers' experiences in our former wars, and what we learned from those conflicts," he responded. "And what's interesting is that, probably starting after the Korean War, the way we trained changed. When you went to the ranges, the targets were no longer bullseyes, they had become human silhouettes, both on marksmanship and live-fire ranges. Counting West Point and active-duty service, I'd been in uniform, then (2003), for about nine-years, and so looking back, now, on that first actual close-up event with the enemy, frankly, it felt just like training. It didn't feel like anything I hadn't done before. Yeah, there were bullets coming at me, and that was a little harrowing (!), but you're still able to focus on what you're supposed to focus on, because of the comprehensiveness of that training. And so, at no point was I, or anyone on the team, shaking in our boots. Everyone recognized, hey, here's what I'm supposed to do. This feels so familiar: Aim the rifle at the target, squeeze the trigger, look for the next target. It very much felt that way."

But what about following the fight? "It's the thinking about it afterwards, where you realize, Oh, S—T!! Remembering that two RPGs had landed just five feet away from me, and luckily the shrapnel blew backwards, and not laterally, because otherwise, I could've gotten peppered with it." Fortunately, in that instance, the armored vehicle he was in helped to keep Squires and his crew from sustaining any injuries. Summarizing with more than a touch of relief and sarcasm: "That was an interesting experience!! All kinds of grenades, and a lot of heavy ordinances were continually coming in on that one, but at the time, you don't think anything about it," said Squires.

There was another enemy incident that stood out in Squires' mind, as it still does. "This one was a tough day. As I'd mentioned, our compound got shelled every 24 to 72-hours. And usually, it was very sporadic. That day, I was enjoying a run with my good buddy, the battalion S-2. A couple of rounds came in, but they didn't seem to be anywhere near us, so we continued our run. But this time it became different. Every once in a while, they'd bring some heat, and this time they did! They must have shelled us with a couple of dozen rockets and mortars. Normally you hear a round or two land, and you can continue with your run, not a big deal. But this time, we quickly realized that we needed to seek some cover in a little bit of a ditch. When we didn't hear any more rounds, we sprinted back to the battalion, back where everyone was. Still hearing no more, we got the sense that it was over," said Squires.

At that point, he and others began to look around the compound for damage, and there was a lot of it. "The worst thing was that there was an engineer unit that was just about to leave the 'wire'," he said. "They were inside the 'wire', but outside of their building, doing the brief before initiating their large convoy. All the vehicles were lined up ready to go, and everyone was assembled up around the hood of the lead vehicle, talking through the final checks for the mission. All of the last-minute details (security assignments, etc.) you do before you go. It happened to be really hot out that day, so all of them had their helmets off, and some had taken their body armor off, as well. When those rounds came in, a couple of them landed right where the group was standing. We ended up losing twelve soldiers on the base that day, with another twenty of them significantly injured," remembered Squires, vividly. By luck of fate, had those rockets come in 10-minutes earlier, or 10-minutes later, chances are good the soldier loss that day would have been half or less, since either the unit members wouldn't have yet assembled out in the open, or the convoy would've already moved out.

And another especially sad memory from that painful day. "From my battalion, that was how the Headquarters and Headquarters Company (HHC) commander, Captain John E. Tipton, died. He was literally standing there, and took some metal shrapnel to his head, and passed away that night from internal bleeding. That was a huge blow. The battalion HHC commander is typically the second command, which is given to the best of the company commanders. He'd previously commanded Alpha Company, and, right before the deployment, took command of HHC. He was just a very experienced and good-natured guy. And so that was a tough one for the officers in the unit to swallow, because we were all so close with him. And, heck, he was my boss. As a staff officer, I was in John's HHC when he died during that mortar and rocket attack (02 May 2004). So that's one memory that doesn't stand out as a good one. But it's definitely a reality-of-war reminder," Squires concluded.

Captain Squires finished his Iraq tour in September of 2004, returning to Fort Riley to continue his company command. With upwards of half of his unit moving on to other assignments, Squires had several months of training-in the replacements ahead of him. His unit had been alerted that it would be returning to Iraq, but those orders would change several times over the ensuing months. As he prepared to relinquish company command and move to Hunter Army Airfield in Savannah (April 2006), his former company did finally get its deployment orders, not to Iraq, but to

Africa!

Squires was assigned to the First Ranger Battalion (1/75th), commanded by then-LTC Richard (Rich) Clark, under whom he had previously served in Italy. Through assignments of increasing responsibilities and promotions to match, as of this writing, now-Lieutenant-General Clark has been nominated to head the U.S. Special Operations Command in Tampa, following his previous assignment as Director of Strategic Plans and Policy for the Pentagon's Joint Staff. That office provides strategic direction, policy guidance, and planning focus, in order to develop and execute the National Military Strategy. Which, then, enables the Chairman of the Joint Chiefs of Staff to provide military advice to the President, the Secretary of Defense, and the National Security Council.

With the Rangers, Squires became an Assistant S-3, and also served as the Rear-Detachment Commander here during a brief overseas deployment for the battalion. After several months with the 1/75, he deployed to Eastern Afghanistan, but in a role outside the battalion. Now in a joint assignment, as the director of an intelligence fusion cell. It was his job to help develop target 'packages,' that could then be assigned to both conventional and unconventional units who would then take the military action needed to eliminate the targeted enemy. "It was a great mission," said Squires. "I learned a ton about intelligence, and I learned how to use it." But thinking back, he recalled that, while it was very easy to feed targets to the elite unit on the base, and get them to take action, it was much harder to pass-off action requests to the conventional units, targeting the "lower- level thugs," despite knowing who and where they were, and having compiled strong intel-based 'packages' on them.

Squires acknowledged that part of their reluctance might have been due to his not being persuasive, or influential, enough in 'selling' the need to get rid of those targets. The other, bigger problem, he concluded in hindsight, was that their goals and his had apparently been different. "I just wanted to see those guys off the battlefield. I had a deck of a hundred bad guys that I wanted to get rid of. But I think they (conventional units) had a greater appreciation for what was going on among the locals, perhaps feeling that if they leave those guys alone, it would be better for community relations overall. They might well have had a better sense, than perhaps I did, of the disruption those operations against the more minor players would cause," he said. Regardless, for whatever the reasons, that don't rock the boat reluctance to take action clearly remained a source of frustration for him.

*Captain Squires leads his Ranger Unit.*

Still

with the 1/75$^{th}$, he was supposed to return from Afghanistan with the unit after three-months, but instead, his deployment lasted closer to six-months, twice the time planned. That extension did not go unnoticed by his wife Sarah, disappointed by seeing the other Ranger unit members return, but her husband not with them! After finally returning to Hunter and family, in October of 2007, Captain Squires helped 'stand-up' the fourth company in his battalion (Delta), with a core group of NCOs. On the next deployment for the battalion, he was assigned to remain at Hunter to finish the critical wartime task of building and training this newest company in the entire Ranger Regiment. "I was dual-hatted as the Delta Company commander, and the battalion's rear-detachment commander, now at the new rank of Major. With both those responsibilities, it proved to be a very busy time," he recalled. Squires used the three-months the remainder of the battalion was overseas blending in, and train up, the almost fifty new-to-the-unit soldiers fresh from Ranger assessment training. Then, after taking his company, along with the others, now returned, out to the National Training Center for combat certification, it was time for the entire battalion team to deploy once again. About a third of the 1/75<sup>th</sup> went to Eastern Afghanistan, including Squires' Delta Company, which ended up being spread between two different sites there.

Experience and training repetition were the keys, in Squires' mind, to the proficiency achieved by his team during this deployment. "You look at the standard Ranger NCO, who was just an unbelievable athlete, trainer, thinker, and leader, and I had a whole company of them!" remembered Squires with admiration and respect. "What we were able to get done, and the proficiency level that we were able to get those Ranger privates to reach, looking back, I know it was because of the quality of those NCO's," he concluded.

Squires reviewed in his mind his troop experiences to that career point. How impressed he was with the U.S. soldier and NCO capabilities he witnessed and experienced early on in Italy. To then having that very positive impression initially deflated upon his arrival at Fort Riley, but restored again when the team deployed overseas to Iraq and rose to the performance standard he felt was necessary for the mission. And then witnessing the performance level go significantly higher with the Ranger NCO's that he had with this latest deployment to Afghanistan. "When I look back at that time with the Rangers, just how strong and capable a team they were because of those Ranger NCO's. They were the primary reason. And then I had a strong crop of officers who could think, plan, and lead, as well, so it all kinda meshed. It was a good four-month rotation. A good mix of operations, with a night off once in a while, to get in some sleep, weightlifting, and planning, and then right back after it." said Squires.

Among the many operations that Squires and his team carried out during this deployment, one really sticks out in his mind to this day. With a group of about sixty Rangers, Squires and his team had flown in at night, via Army Chinooks, and purposely landed about six kilometers away from their intended target. And the target for that mission was an individual insurgent who had been identified, his movements watched, and his location established. Their intelligence was solid.

As they were walking in toward that target area, Squires heard a single gunshot. "I thought back to my time in Ramadi and remembered that no one just shoots a single shot in the middle of the night. That's a signal. That's someone warning someone else that we are on the ground. So, I sent out a radio call over the net. Hey, be prepared. Something's not right. Keep extra scrutiny. And sure enough, in less than two minutes, all hell broke loose," he remembered vividly. His group had suddenly come upon an entrenched enemy force! "That was a pretty harrowing experience," Squires recalled.

Made so, because the enemy had the high ground, while his team of soldiers had walked down below. In hindsight, they thought they'd picked a safe place to land for the target approach. Unbeknownst to them, it was the classic wrong place, wrong time. "The area wasn't as safe as we thought it was. They had two heavy machine guns entrenched on a hilltop, and we sustained two casualties pretty quickly. One of our platoon sergeants took a round through the side of his body armor and into his chest. And for another Ranger, the same thing, a round to the chest. So right away, I had two significant casualties on my hands, and an enemy that I could not dislodge. Nor was it my mission to go up and dislodge them. That would've been a huge risk for no gain," said Squires.

He was then faced with the difficult task of extracting his soldiers, two of whom were seriously wounded, as his Rangers remained under heavy enemy fire. Although there was an Air Force F-15 on-station well above them, it was not safe to attempt an assist. "Because we were so close (to the enemy), it didn't have the ability to do anything other than shoot 20-mm rounds, and those rounds are designed to shoot down other airplanes, not to shoot bunkers," said Squires. "We were too close to that hill-top and pinned down at that point with no cover to move to. Maybe that F-15 could've done some work, if I could've backed off another 500-meters, and he could've dropped a 500-lb bomb, but I didn't have that luxury at the time," he recalled.

His tense situation under fire stalemated until about 45-minutes later, when an AC-130 gunship finally arrived high above the fight. At that point, with the AC-130's lethal, pin-point accuracy of fire, that unique, situation-dominating flying gun platform completely eliminated the hill-top enemy, and with that, the battle was over. "Once that asset (AC-130) showed up, and fired a couple of rounds, I was no longer concerned, I knew we'd be fine," said Squires. "That gave us the ability to move the casualties and extract the whole ground force a couple of kilometers away to a clearing, and then call in some assets to pull us ought of there. Obviously, we had to scrub that mission. Which was really unfortunate because we had such good intel on the target we were heading to, and we had flown in a pretty robust ground force." he said.

Squires took great pride in the team (Delta Company) that he had stood up, molded, and led on that Afghan mission and others. In his view, undertaking missions with Rangers made a real difference in effectiveness, and in the assets that could be called upon to assist them, such as, in this case, the F-15, and especially the AC-130, which ended the extremely dangerous standoff, in a hostile setting, by destroying the enemy fighters. "There's no doubt that we performed as well as we did because it was a Ranger unit," said Squires. He went on to praise the expertise of Ranger medics, the very best in the Army, in his experienced view. Faced with the early casualties during that mission, "one of the doctors (medic), bullets still flying, crawled forward, hooked his safety line to one of the wounded Rangers, and literally dragged him out of the direct combat zone to safety," so that he could more effectively treat him. That, to Squires, stood out in his memory as an extremely courageous act. And thanks to the rapid, on-site treatment by the two expert Ranger medics there on the ground, and follow-up surgical care once flown back to safety, Squires' two severely wounded soldiers both survived, although both did face, as might be expected, a very long recovery.

Once back at the compound, and seeing after his men, accounting for the time difference back in the states, Squires called his wife in Savannah. She was hosting a ladies social gathering at the time of his call. After a bit of 'how's your day going, I was in a firefight' conversation, Sarah Squires sensed there was more to this call than normal. Her Army wife instincts kicked in. She concluded this wasn't just a regular call home. Without him saying so, she knew he'd been injured during the mission he'd told her about. She was right, of course. Squires <u>had</u> been hit! He'd taken some shrapnel in one elbow from a nearby enemy RPG which, fortunately, hit and blew by him, not at him. Nothing serious, not to worry, he'd be fine, hadn't missed a step, he assured her. Regardless, he had been wounded in combat, and a Purple Heart medal would follow. Aside from the obvious, there's a lesson to be learned here. Over time, most Army wives develop an infallible internal BS meter. Thus, even by the sin of omission, they cannot be fooled. Best not to try.

Squires returned from Afghanistan, shortly after that intense nighttime mission, and turned over Delta Company to another officer on the staff. He had completed a total of 23-months in combat (Iraq/Afghanistan), and during two additional unit deployments, had remained at Hunter, serving as the Rear-Detachment Commander. After his final Afghan tour, Squires then left the 1/75th, having been selected to attend the Command and General Staff College at Fort Leavenworth, Kansas (2009-10).

This year was one that promised to be enriching, both from a schooling perspective, and for the opportunity to re-connect with his family, after having spent so much training and deployment time away. From a personal standpoint, however, it proved to be a year of unexpected challenges.

Prior to the start of his studies, the Squires family scheduled some vacation time, flying to Tacoma, Washington to visit his wife's family. While there, he, Sarah, and some visiting relatives decided to do the scenic, six-mile+ 'Sound of Narrows' run. That was a Saturday, enjoyed by all. The next day, his wife's folks put together a nice welcoming gathering for the Squires family and the other relatives. During that event, Squires remembers not feeling well. He was having a hard time breathing normally, somewhat like what a chest cold can do. With another week of vacation ahead of them, his wife suggested he go to the medical facility at nearby Fort Lewis to get looked at. They did a normal work-up, gave him some antibiotics and he returned to his in-law's home.

Asked how he felt the next day, he told his wife that when he breathed in, he'd get a spike of pain in his left shoulder, and similar pain in his right ribs. And the deeper he breathed in, the worse the pain became. He concluded that he must have pulled something on that Saturday run. Two days running, he woke up bathed in sweat. His wife told him it was time to go back to the docs at Fort Lewis. Based on his symptoms, they took some blood along with an intensive series of tests. Then, as the day wore on, more tests, and still more blood drawn! Finally, after enough time to think and worry about what's going on with his health, quite natural for all of us, a Lieutenant-Colonel came in, and after doing an unexpected, but thankfully, quick exam of his rear quarters, the Doctor informed Major Squires that has had no internal bleeding, (the good news), but what he did have was a pulmonary embolism (the not so good news). Exam over, that announcement made, the Doctor left.

Fortunately, before long, a doctor from the Internal Medicine Department arrived and explained in depth what Squires was experiencing. This is a common malady, he was told, where folks from overseas are in good shape, fly back to the States, and become sedentary for a while on that long flight home, which then allows clots form in their legs and either not break off, or in fact, eventually break off. "Although I never felt any kind of DVT (Deep Vein Thrombosis), something certainly broke off, some kind of clot, and got stuck in my lungs," remembered Squires. Had the clot become lodged in his heart, instead of his lung, he likely would not have survived (In fact, Mike's older brother, Neal, had passed away from a clot in his heart just two months prior). The doctors were actually amazed that he was still alive. Once he was given blood thinners, he began to feel much better. It also surprised the doctors that he didn't appear to be having any residual issues with any of his limbs. "I bounced back, and literally a day later, I was fine," he recalled.

Squires then contacted the 75th Regimental doctors to explain his situation. They reassured him that this problem happens frequently. He was prescribed blood thinners for the next several months to reduce the possibility of additional clots forming. Following that, there were additional medical tests to take, all of which he passed with, as the saying goes, flying colors. He returns to Leavenworth. The Command and General Staff courses go well. "I try out to come back to the Ranger Regiment. I get hired. Supposed to go to the 3rd Ranger Battalion at Fort Benning.

And things are looking good. I've got the next six-to-nine years mapped out, through the various probable assignment progressions, building toward potential Ranger battalion command," he recalled.

Major Squires is still progressing through the coursework at Leavenworth. After six-months on blood thinners, then off them for a month, he takes some final tests required to establish his health status, at a large medical facility in nearby Kansas City. In February, he got a call from his doctor there, indicating that he needed to come in for a consultation. Once there, it was explained to him that he had a condition called Protein-C Deficiency. Protein-C is a critically important element since it works to prevent clots from forming within the body. Squires quantity and quality of Protein-C was markedly deficient. Therefore, to avoid the certainty of future clots, which would likely be fatal, his doctor explained that it would be necessary for him to take blood thinners for the remainder of his life.

"So now, I'm devastated, because I know what that means," recalled Squires. "They (doctors) didn't realize that my six-to-nine-year future career in the Army has just been completely derailed. That was a really hard time for Sarah and me, both the stress leading up to those tests, the outcome, and then not really knowing what we were going to do after that," he said.

So, what should have been an enjoyable year with family at Fort Leavenworth (2009-10), became one of career uncertainty, impacting both Squires and his wife. He knew, without question, that he would no longer be able to join an operational (warfighting) unit. An understandingly tough reality to face for a bona-fide, seasoned warrior. So, he would be leaving the 'operating' side of the Army, to experience additional professional growth assignments within the Army's 'generating' force. But if there was one positive, throughout all of this personal turmoil, his health crisis did bring him and Sarah even closer together, as they dealt with it all together as a couple. And another definite positive: Despite the health issue, he attends the prescribed CGSC classes, completed all the required coursework throughout that year (2009-10), and graduated on schedule. Fortunately, there would then be many attractive career opportunities ahead.

Moving forward, Major Squires placed a call to a long-time friend, the manager at the infantry branch personnel office. They had previously discussed the fact that, pre-diagnosis, he had been slated for a return to the Rangers at 3/75 at Fort Benning, but pending test results, he might be in need of an alternate job assignment. And that became the case. He knew, then, that he wouldn't be going to 3/75. So, Squires requested that, if possible, he be sent, instead, to the Ranger Training Brigade. And it worked out. His request was granted. "So, I got to go down to Florida and become the Operations Officer at the Ranger 6th Training Battalion, which was just a fantastic job! Doing a great mission with a great crop of folks," he recalled.

That turned out to be a terrific year (2010-11) and experience for Major Squires. The following year (2011-12), he became the brigade-level Operations Officer for the entire Ranger Training Brigade. "With a great boss, and a great group of people, that was a really good experience, as well."

Looking back to the point of that career-altering diagnosis, as soon as it had been determined that, because of the continual blood-thinner requirement, he would no longer be able to be with fighting unit, he knew in his heart that he wanted to go teach at his alma mater, West Point. "I knew I could make an impact, by helping to lead that next generation of officers. So I put in a packet to teach at West Point, and got hired to teach in the economics department there. Which is what led me to (first) go to M.I.T. for a couple of years to get a master's in business administration," said Squires.

Because of his selection to join the West Point faculty, the Army, then, selected him to earn his MBA degree (2012-14) prior to beginning his instructional duties. He found the two years at M.I.T.'s Sloan School to be some of the most intellectually stimulating of his career to that point. Extremely competent faculty, coupled with his degree-class of very bright fellow students, made for a truly enriching personal experience. It was during this time that Major Squires was promoted to Lieutenant-Colonel (2013). Upon receiving his MBA, as planned, Squires then taught economics at West Point for a year (2014-15), prior to coming to Hunter Army Airfield in Savannah, GA (2015-17) as its Garrison Commander.

Although the lifelong blood thinner regimen required that he shift his Army career track from operational assignments to helping generate future soldier-leaders, his overall health remains excellent. He still runs, does daily PT, and continues to meet or exceed all annual Army physical fitness standards and expectations.

Following his two-year term as Hunter Army Airfield Garrison Commander, in the Summer of 2017, LTC Squires reported back to his alma mater, West Point, this time to serve a one-year assignment as a Regimental Tactical Officer (RTO).

## PERSPECTIVES ON LEADERSHIP

Concluding our interview, LTC Squires was asked his thoughts or experiences regarding certain elements that make up the profession of soldering. To begin, what memories stood out in his mind about the individual courage he has witnessed? Without hesitation, he recounted the example covered earlier of the Ranger medic who ignored enemy gunfire and RPG's to crawl forward to retrieve a badly injured fellow Ranger during a nighttime firefight in Afghanistan. And then one other example of courage, this one collective, came to his mind. "It's always impressive to watch the first guy go into any door (in a combat zone), because you never know what's behind it. Not as bad for the second or third guy, because the first guy through is gonna keep whoever might be there occupied. Courageous and always impressive," he said.

An example of leadership, personally witnessed? "It depends on the level. There's one I distinctly remember at the squad level, being on a target with the Rangers in Afghanistan. Some enemy they were going after 'squirted' (got away). They got outside of our perimeter. And without missing a beat, one of those fire-team leaders is like, "Hey, follow me"! And he takes off up the mountain chasing after them. That's outstanding leadership at the direct level. Leadership at that level is much easier to see. Much harder to detect at the organizational and strategic level, because it's so much more nuanced," said Squires.

Are solid, great leaders born or made? "I've thought about this a lot," he responded. "The Army's answer is that it's teachable. We'll take whoever you are, and we'll give you experiences, and some leadership training, and you can be a great leader. I'm in the Army. It's my duty to believe that. However, everyone I've seen do well in the Army just seems to have a little something there that I think is hard to teach. Some kind of ability to command a room. And I've never seen anyone learn that. For instance, I've never seen someone at the lieutenant level, who didn't have it, and somehow figured it out and gained it as a captain, or major or colonel. So it's hard for me to say that there's not something innate there that's very intangible and hard to define. I have seen people who were poor (leaders) get better, but I haven't seen anyone go from poor to good, then to fantastic, someone who can really command a room. But when you think about it, we do generate leaders all the time, starting with privates who grow with experience and training and then put in front of some troops, and those guys figure it out," concluded Squires.

What is the best life-lesson you've learned so far in your career? What has the military meant to you? "The one word I think of when I think of my time in the Army, and what it's done, is grit. It's taught me that life's gonna give you a bloody nose. Maybe that's a pulmonary embolism, maybe that's two guys badly injured with chest wounds in combat, maybe that's some shrapnel to the elbow during a mission that turned crappy, but you get up from being punched in the face and you solve whatever problem life's put in front of you. There's no doubt that the Army has taught me that time and time again. And I'm better for it," said Squires.

LTC Squires then further summarized his time in the Army, and as Garrison Commander. "So, it's really been interesting," he said. "Although I'm in the Army, I've had, it seems, five Army careers. Beginning with the conventional career, then special operations with the Rangers, then training assignments, followed by academic career, and now this garrison role. And I've learned something different and valuable at each stage. I'm very thankful for all those experiences I've had, because all of them have contributed to preparing me for this particular job. Garrison command has been very enjoyable, because I've been able to reach back and literally touch all of those previous eras and phases of my career, and use those skill sets in this job. Maybe that's in a teaching role, like I do at Rotary teaching community businesspeople more about the Army, or in a negotiation-influence role with folks in local government, or in the partnership and communications things that we do. All of it's come together in this Garrison Command billet. It's been fantastic," he said with both genuine feelings and satisfaction.

Following his very successful tenure as Hunter AAF's Garrison Commander, Lieutenant Colonel Squires now moves back to West Point, this time to become a Regimental Tactical Officer in charge of one of the school's four regiments.
"I'll be working with cadet leadership, military training, discipline, standards, and physical

*LTC Mike Squires*

fitness, all of those facets that comprise what we need an Army officer to be," he said. Having been on the academic side there previously, he's looking forward to experiencing the other side of it, the overall, comprehensive molding of those future officers.

Our very best wishes to Lieutenant Colonel Mike Squires, wife Sarah, and family, as together they move forward to the very well-earned and deserved future opportunities that will continue to be theirs.

**UPDATE**: (Summer, 2019) Recalling that LTC Squires had been diagnosed with an ailment which, although easily controlled with medication, led Army leadership to determine that his status would need to become classified as "non-deployable." This was during a period when the top levels of the Army, uniformed and civilian, had established a policy that those deemed non-deployable would eventually have to be released from further service. So, despite the fact that Mike Squires was a dedicated, well-educated, multi-skilled, proven leader in combat, as well as a West Point faculty member, and as a very successful garrison administrator (Hunter AAF), and, thus, very well qualified for any number of non-combat Army positions, and despite the support of several General Officers familiar with Mike's capabilities, the Army leadership made the decision that LTC

Squires would be medically-separated (Spring, 2019) from the military branch and service that he loved. Prior to that, his West Point assignment, originally scheduled to last for just one-year, fortunately had continued for a second year, during which time, as a Regimental Tactical Officer, he was then promoted to take charge of the cadets in all four of the Academy's regiments. There was, however, a very important move from the Army, as Mike began what would be his final year on the staff at West Point (2018-2019). With Army retirement now ahead of him, he received the well-deserved promotion to Full Colonel to serve out that final year at the Academy, and this time as the Brigade Tactical Officer.

**UPDATE:** (Summer 2024) As often happens with well-qualified and highly thought of former military officers, upon completion of his Army service (2019), Colonel Mike Squires (Ret.) quickly secured a great, career-related position with the Culver Academy, a military high school in Culver, Indiana, where he now continues to serve as the Commandant of Cadets. No question, he has been a very welcome and successful addition to that highly regarded school.

# About the Author

William (Bill) Cathcart holds a Ph.D. and master's degree in mass communication from The Ohio State University and a bachelor's degree in history from Hope College. After ten years as a university professor, he transitioned to commercial television work in Minnesota. Then, in 1985, he bade farewell to his snowblower and welcomed the opportunity to move to Savannah, Ga and WTOC-TV. He retired from WTOC after 29-years of service, his last 24 years as the station's Vice President and General Manager.

Dr. Cathcart has volunteered and served on several nonprofit boards for many years, including the Savannah-Area Chamber of Commerce, Chair of the Military Affairs Council, and the National Museum of the Mighty Eighth Air Force Museum in nearby Pooler, GA. He is also a longtime member of the Rotary Club of Savannah.

As a result of his decades of demonstrated commitment to our military, he has been granted honorary member status with Hunter Army Airfield's 1st of the 75th Rangers; Hunter's 3rd of the 160th Night Stalkers; and the 3rd Infantry Division's Combat Aviation Brigade (Hunter). He has also been named a Third Infantry Division "Rock Star" ("Rock of the Marne") at Fort Stewart, Georgia. Additionally, he served for many years as the Honorary Wing Commander of the Georgia Air National Guard's 165th Airlift Wing in Savannah, Ga.

In 2013, Cathcart was nominated by then-3rd Infantry Division Commander, Major General Robert B. "Abe" Abrams, and selected by the Department of the Army, to become the first Civilian Aide to the Secretary of the Army (CASA) for the Coastal Georgia Region. In 2023, after serving for the maximum ten-year term, Cathcart was granted official CASA Emeritus status by the Department of the Army.

Made in the USA
Columbia, SC
28 October 2024

44884875R00172